Ebertin — Hoffmann

FIXED STARS

Published By

AMERICAN FEDERATION OF ASTROLOGERS, INC.

6535 South Rural Road — Tempe, Arizona 85283

Mailing Address

P.O. Box 22040 — Tempe, Arizona 85282

EBERTIN – HOFFMANN

FIXED STARS
AND THEIR INTERPRETATION

Translated by

Irmgard Banks, Melbourne, Australia

EBERTIN – VERLAG, D 7080 AALEN

The title of the German edition is:

"Die Bedeutung der Fixsterne".

1st English Edition 1971
2nd English Edition 1973
3rd English Edition 1976

INDEX

INDEX, continued

Preface

For many years, numerous friends of our science have expressed their wish to have a working basis which incorporated fixed stars into natal charts and, as well, to have a review of a cosmic analogy to world events. Books treating the subject of fixed stars extensively were first published several decades ago. They are now out of print and have not been re-published. Most books of this kind took, as their basis, tradition, and neither took into consideration later research nor amplified research with verificable examples. However, there was published a little book by ELSBETH EBERTIN in 1928. The title was "Sternenwandel und Weltgeschehen" (1). It included cases of her own records with notations regarding her own investigations. Now, we have decided to enlarge this booklet and make it richer with more and newer cases. The basis of this new book is therefore ELSBETH EBERTIN's data. It is our intention to bring back to memory collected notes of the investigations of this genuine researcher. In the 11/12. vol. of the journal "Mensch im All" (3), today published as "Kosmobiologie", new cosmobiological research has been used in a detailed treatise written by REINHOLD EBERTIN using the name of his mother. Another co-worker of ELSBETH EBERTIN (in the second decade if this century) was GEORG HOFFMANN, who has now taken on the task of editing this text, using new sources of research in it. REINHOLD EBERTIN added many cases when finally revising this text. Therefore, "Fixed Stars and their Interpretation" is very much a communal effort, with the contributors being the late ELSBETH EBERTIN, GEORG HOFFMANN and REINHOLD EBERTIN. Many additions and corrections have been made by Dr. WALTER KOCH. He specially endeavoured to explain fixed star names and their position in constellation.

The task of re-calculating the positions of the fixed stars was abandoned, as we had at our disposal several tables, those of Dr. WALTER KOCH, Goeppingen (4), R. SCHUMACHER, Munich, ERICH VON BECKERATH, Wiesbaden (5), GUSTAV SCHWICKERT, Graz (+), and Dr. TH. LANDSCHEIDT (6). The reader therefore should not find it difficult to calculate intermediate values.

The source material for this book grew to very large proportions. On the recommendation of Dr. WALTER KOCH, several fixed stars were not included in this book, as they were either of minor importance or the material relating to them was considered too vague.

Some History of Fixed Star Interpretation

The observation of fixed stars and their relationship to world events had already begun thousands of years before our time. By analogy to the yearly motion of the Sun, an allotment of 360° was made to the circle. By the 6th century B. C., the Babylonians had a heavenly large circle system, incorporating Meridian, Horizon, Equator and Ecliptic. Dr. W. KOCH wrote in "Kosmobiologie" (7) a narration "Zur Fruehgeschichte des Horoskops" (= "Early History of the Horoscope"). Measurements were taken inside this system and this was further portrayed by way of a celestical firmament globe, the latter having been introduced by Greek astronomy. The Babylonian "Mul-A Pin" reiterated 16 constellations positioned on the path of the Sun (this in the 8th century B. C.) and about another 70 contained in other parts of the heavens.

The first fixed star catalogue was compiled by TIMOCHARIS and ARISTYLL about 250 B.C. in Greece. HIPPARCH about 150 B.C. improved it, and it was handed on to future generations in canonical form by PTOLEMY in his astronomical hand-book "Almagest". Two examples may serve to show how exact PTOLEMY was in his work. In the original text of PTOLEMY regarding Regulus, Alpha Leonis, was read: "Leo 8. The one in the heart of the so-called Basiliskos, longitude 2°30' Leo, latitude + 0°10', magnitude 1" and regarding Spica (Wheatsheaf), Alpha Virginis: "Virgo 26°40', latitude - 2°, magnitude 1."

The names of fixed stars have their origin in the Babylonian era. As seen in the Gilgamesch-Epos, only eleven signs were distributed to the ecliptic. Libra then was still considered the "scissors" of Scorpio. There certainly was not a uniform distribution of the signs. For the month Ijar (April, to be correct April-May) there was, not only Taurus but also the Plejades, and for the month Sivan (May-June) there was not only Gemini but also Orion.

Originally the fixed stars were associated with weather phenomena. E.g. Sirius, rising early in July (Northern Hemisphere that is) is supposed to give rise to the feverish heat of the "Dog Days" (8, 9).[x]

---

[x] Footnote: a name given to hot days in the Northern July summer heat.

The effect of a fixed star on the weather was not its relationship to other stellar bodies, but its position when rising and setting. PTOLEMY did not place particular importance in these "star phases", but he concluded that the atmospheric phenomena are influenced by the planets. Finally 15 stars of first magnitude and 15 stars of second magnitude were chosen and used for special observation. Gradually the influence of fixed stars was brought into association with the body in health and the body in illness.

The symbols of the fixed stars served mainly as originators of their meaning. For example, it was believed, that a man with an accentuated position in his horoscope of The Baker, would like those parts of the country which abounded in water, and would become a vine grower or would enjoy drinking wine. To The Whale were ascribed fishermen, and also "drinkers and gluttons, prison warders, robbers, occupations connected with salt pickling, storekeepers, misers and dishonest keepers of the seal".

Just as certain planets were assigned to the various signs, eventually some planets were co-ordinated to their decanates (allotment of 10 degrees each), and thus the effect of fixed stars was compared with other stellar bodies and these were then used to describe the character of the fixed star in question. CLAUDIUS PTOLEMAEUS (10) lived about 100 to 180 A.D. He was a renowned mathematician, astronomer and astrologer, and the founder of the world system named after him, the Ptolemaic system. A paragraph (1, 8) "About the Effects of the Fixed Stars" in the "Tetrabiblos", written by him, will be shown here, - in the wording of M.E. WINKEL - who added the Arabic version to the names of stars. However, these were wrong or in their antiquated form (Dr. K.).

"Now, at this point, we want to continue to describe the powers of the fixed stars, derived from their nature. I shall line them up so that I fit them to planets according to their similarity.

The stars in the head of the Ram (Alpha Arietis, Beta Arietis) shows a blend of the nature of Mars and of Saturn. However those stars in the mouth of the Ram have a nature similar to Mercury and a little of Saturn.

Those in the hindfoot belong to Mars, in the tail to Venus. There, where the star figure of the Bull appears as cut off, the stars will have a Venus nature; the Plejades however belong to the Moon and to Mars; the one star in the head, between the Hyades, because of its reddish glimmer (Aldebaran), called the beacon, belongs to Mars.

In the star picture of Gemini are located between the feet stellar bodies of the nature of Mercury (Propus, Alhena) and to some degree of Venus, the clearly bright ones in the area of thighs (Wasat) have a Saturn nature. In the heads are situated two bright glowing ones, of which the first one (Castor) has the name Apollon; and the second one, of Mars nature, (Pollux) has the name Herakles.

In the eyes of Cancer are situated two stars of like nature, having an effect simultaneously of Mercury and a little of Mars. Those in the scissors belong to Saturn and to Mercury. The particular spot of mist on the other hand, the one in the chest, called Praesepe, again relates to the Moon and to Mars. Close to it are situated two stellar bodies, called donkeys, and of Mars and Sun nature.

In the jaws of the Lion are two ones belonging to Saturn (Ras el Asad) with a light blend of Mars nature, on the neck three ones to Saturn (Algieba) with a blend of Mercury. The bright shining one in the heart of the Lion, Regulus, to Mars and Jupiter. The two ones in the thighs and the brilliant one in the tail (Denebola) have Saturn and Venus nature. The remaining ones in the shank are Venus and to some degree Mercury nature". In the same way the listing is continued with the relevant equivalents. Nowhere are any cases given.

In our interpretation of fixed stars, we have tried to include many cases, in order to clarify just how there are any "effects" resulting from fixed stars. If the words "effect and influence" are used, this is not meant to indicate a causal relationship between stars and men. But the matter in question is one of synchronisation, analogies and matters relating to each other.

The examples cited are meant to be ones relating to the radical position in the cosmogram. Documentary evidence regarding the analogies of fixed stars and planets and weather will be published at a later date.

For practical use, we would like to point out that only fixed stars in conjunction with other stellar bodies with small orb should be used, if possible one should work with no more than 30'. Perhaps if one had a larger collection of cases, one could also use the opposition.

Some fixed stars are of very great declination from the ecliptic. Some practising cosmobiologists are of the opinion that fixed stars with large declination, and if more than 23° away from the celestial equator, are of no use in practical interpretation. Practice, however, does not bear out this opinion. Moreover, practice shows that this declination does not appear to matter at all.

FIXED STARS AND THEIR INTERPRETATION

1. DENEB KAITOS, 1°51' Aries (1950), Beta Ceti, magnitude 2,

also called Diphda, is situated in "the Tail of the Whale". Its property is that of Saturn. Associated with it are inhibitions and restraint in every way, psychologically and physically.

On 5.5.1945, the end of World War II, Mars was in conjunction with Deneb Kaitos.

The poet HOELDERLIN had a particularly tragic fate. In his natal chart, Sun was in conjunction with this fixed star, and the half sums Moon/Neptune = Neptune/Pluto. The poet, when 30, went catatonic, a kind of Schizophrenia, when Pluto s triggered off the said configuration.

Deneb Kaitos is found conjunct with the Moon in the horoscopes of HEINRICH HIMMLER, HITLER's Gestapo Chief, and CANARIS, Chief of German counter espionage. HIMMLER committed suicide when the situation became hopeless. CANARIS was executed in connection with the plot against HITLER in 1945.

2. ALGENIB, 8°31' Aries, Alpha Pegasi, 1st magnitude,

"in the Wing on the Side of Pegasus" is supposed to have a Mars/ Mercury nature. Therefore it brings about a penetrating mind and a strong will as well as determination, an impressive way of speaking and a gift for oratory. Conjunction with the Sun will make for a fighting spirit and a love of learning. Connected with Mercury, Venus and Jupiter, one can count on popularity and interest in art and literature, and distinction; if not further configurations also point to similar gifts. With Saturn, in general, it is thought of as an impeding factor, but has a name for good memory according to MORIN (10). If angular with Mercury and Uranus, Algenib will make for an inventive spirit.

The conjunction of the fixed star with MC is found with the former Minister of Propaganda, ready-witted speaker, JOSEF GOEBBELS, and also Minister for Central Germany GROTHEWOHL. The conjunction with Mercury is significant in the Italian physicist MARCONI, and the conjunction with Venus in the popular Italian tenor GIGLI.

3. SIRRAH, 13°40' Aries, Alpha Andromedae,

in the "Head of Andromedae" is supposed to have properties of Venus and Jupiter. In good aspects, it is significant for an harmonious nature, which in itself brings about a good relationship to other people and makes for popularity. If it is in affinity to propitious stellar bodies and with the personal points MC, AS, Sun, Moon, one can count on becoming well known in public, and popularity with the masses. The French Novelist EMILE ZOLA, (2.4.1840) had in his natal chart Sun in 13° Aries.

If Sirrah is connected to the Sun, the native can easily become unpopular and can be toppled over. Also, if transiting Saturn passes over the fixed star in conjunction with another stellar body, a weakening of popular appeal is indicated. In this way, E. EBERTIN noted that, in many cases, directors and administrators, who had in their cosmogram Sun in conjunction with Sirrah, lost their posts, when in 1938 and early 1939 transiting Saturn passed over this conjunction.

In BISMARCK's the "Iron Chancellor's" cosmogram, we find Sirrah in conjunction with Sun. He forged the Second Reich, but was then made to resign under the young German Kaiser WILHELM II.

The natal chart of the former Minister for Foreign Affairs, VON RIBBENTROP, shows a conjunction with MC. He experienced a quick rise and then ended on the gallows. The Russian revolutionary LENIN had the fixed star conjunct with Mars. Significant for the eminent inventor EDISON is the conjunction with Uranus = Mercury/Jupiter.

4. BATAN KAITOS, 21°17' Aries, Zeta Ceti,

in "The Whale", has Saturn character. The "Whale" means really "monster". The Saturnine properties, such as inhibition, reserve, caution, solitude, and simplicity often are forced on to such people, either by a mundane or a higher power. Sometimes ideas are propagated which make life for the native trying or troublesome. To such persons, fate is usually one of change. People influenced thus tend to depression or dwell on the thought of death. Life often is full of humiliation and renunciation and obstacles. But the position of the fixed star in the complete chart is always important.

In the natal chart of WILLIAM BOOTH, the founder of the "Salvation Army", the Sun is in conjunction with Batan Kaitos. BOOTH was successful in spite of many humiliations and difficulties, and especially with those converts, the "saturnian" people.

In a criminal, born 9.4.1864, we find the same conjunction with short interruptions only, in gaol and in hard labour. LENIN also had Sun conjunct Batan Kaitos.

General LUDENDORFF, born with Sun in conjunction with this fixed star, had a life full of change: During the 1st. World War he was the celebrated commander-in-chief; emigration after the war, afterwards connected with mass movement and comrade in arms with HITLER, then founder of a kind of religious movement. The same configuration is found with the French politician LEON BLUM, who was Prime Minister several times. A depressed manner and pre-occupation with thoughts of death is shown in the conjunction of Batan Kaitos and the Sun in the stigmatized THERESE VON KONNERSREUTH.

5. MIRACH, 29°46' Aries, Beta Andromedae,

the "Side of Andromeda", corresponds to the nature of Venus with a Neptunian influence in the positive sense: cheerfulness, happiness, love of company, and many sided interests, but also altruism, tendency to inspiration and medium-ship as a base for artistic creation. These people have a stimulating effect on others, make friends easily and are helped on in life by others. This is especially true in aspects with Venus, Jupiter and Sun, with a well placed Neptune, and in connection to MC and AS. However, should the whole cosmic picture be an unfortunate one, then these helpful configuration with the fixed star are not of much effect.

The influence of Neptune and Venus was easily seen in HEINRICH NUESSLEIN. This man had the gift of medium-ship and in trance painted several thousand pictures. On the other hand, he was a man who liked good company and enjoyed happy times. The conjunction of the Sun with this fixed star is seen in the natal chart of the composer MAX VON SCHILLING, who became known by his opera "Mona Lisa" and the melodrama "Hexenlied" (= song of a witch). We find a conjunction with the Moon in the chart of the composer, VERDI whose operas still have a heart wrenching effect on the soul (Moon) of the listener. When the English Prime Minister CHURCHILL was born, Mirach was in conjunction with Neptune.

6. MIRA, 0°50' Taurus, Omikrow Ceti,

stands for "Stellar Mira", the "marvellous star" in the "Tail of the Whale". It is a double star, given this name because of its changing lights, which go from 2 to 10 in variation of light in 332 days. Its nature corresponds to Saturn and Jupiter. Connected well and with a well placed Saturn, and linked up well otherwise, this will mean prudence, perseverance, versatility, a progressive spirit and endurance in solving difficult problems. This fixed star in unhelpful connections will bring as results failures, fiascos, enmity, especially with Saturn transits and when linked with the Moon melancholy will then appear. Linked with Mercury, it will mean a lessening of spiritual forces. In critical situations, suicide is indicated.

In the natal chart of NAPOLEON III, born 20. 4. 1808 in Paris, the Sun was in conjunction with Mira. This particular NAPOLEON had changing fortunes. First he followed his mother into exile, later he was student in the Augsburg Gymnasium. In 1836, he led a rebellion in Boulogne and Strassbourg. Then, for the first time, he was sent by naval vessel to America. Then he was sentenced to life imprisonment. In 1846, he succeeded in escaping from the

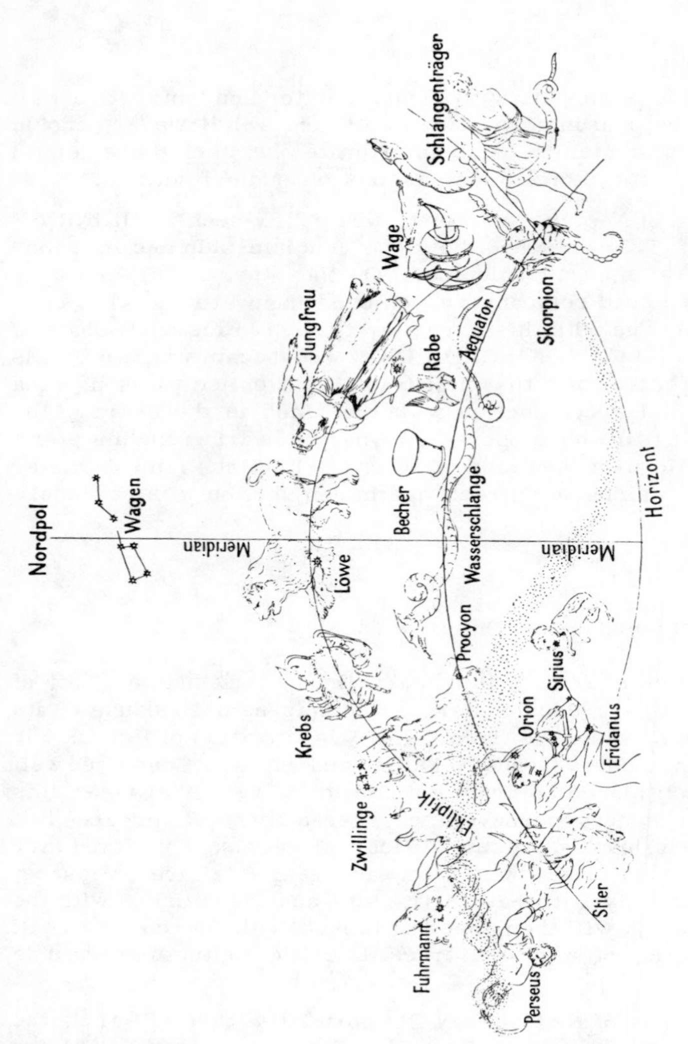

Fig. 1: Segment of the Zodiac in the evening in Spring. The sign Taurus is just setting and above it the constellation of Perseus who is a symbol of Mithras, and according to legend, he is putting The Bull to death. Culminating is the constellation Leo with the fixed star Regulus, Scorpio (on the right) is rising. The fixed stars are marked in the various constellations.

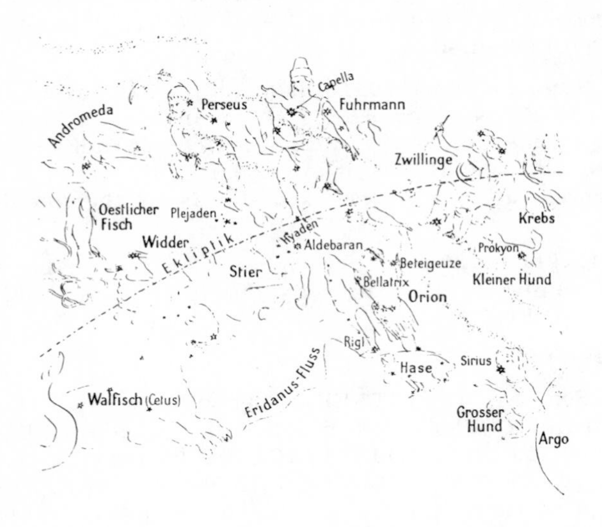

Fig. 2: Section from the "Heavenly Picture Book" from Aries to Cancer. The most important fixed stars are marked. The latter are measured in astronomical longitude along the ecliptic, although they have quite a different distance (latitude). Deneb Kaitos "in the Tail of the Whale" is situated near the ecliptic. Sirrah "in the Head of Andromeda" shows a larger distance - and is not visible in this figure. Almost in the same direction are Algol "in the Head of Gorgo Medusae" und "Perseus" in the Plejades in the last degrees of "Taurus". The sign Gemini is at further distance but one can see the beginning of it with the "Northern Bull Eye" and "the right Eye of the Bull" a little further off the "Hyades" and not far from "Aldebaran". Rigel is situated in the left knee of "Orion" somewhere further down (Southern Latitude). Also in "Orion", but on the "Left Shoulder" is "Bellatrix" and on the "Right Shoulder" Betelgeuze will be seen. "Capella" is quite marked in "The Charioteer". From these examples can be seen how in antiquity man could find his way round on the starred sky by the simple method of putting together into constellations the most important stars.

fortress Ham. In 1848 he was made President of the French Republic by vote of the people, and in 1852 he was crowned Emperor of France. After the battle of Sedan, he was sent as a prisoner to Castle Wilhelmshoehe. In 1870, France became a Republic again, and in 1873 NAPOLEON died in England.

King KARL I of Rumania, born 20.4.1839, did not have a similar fate. He could and did help his people to advance, and in 1881 Rumania became a Kingdom. KARL did not suffer set backs of the kind NAPOLEON did. One always has to consider the entire cosmogram, the configuration with the fixed star is not the sole determining factor.

A boy, born 21.4.1911, 8.30 p.m., when carrying another boy on his shoulders while playing, fell so badly that he broke his neck. (Taurus rules the neck!)

Further examples:

Conjunction with Sun: ADOLF HITLER, Chancellor of the "Third Reich", Writer PERYT SHOU (Peter Schulze), Queen ELIZABETH II of England, LEONARDO DA VINCI, the Italian painter.
Conjunction with Moon: MAXIMILIAN DE ROBESPIERRE, who introduced a reign of terror after the French Revolution and had many of his adversaries executed, then himself died by the guillotine. The miracle healer GROENING had tremendous success, but also made enemies.
Conjunction with Ascendant: Colonel REDL, Austria, (high treason) died by suicide.
Conjunction with Jupiter: The actress GERTRUD KUECKELMANN.
Conjunction with Neptune: Poet RAINER MARIA RILKE.
Conjunction with Mercury: Poet JOSEF AUGUST LUX.

7. EL-SCHERATAIN, 3°17' Taurus, Beta Arietis,

in close conjunction with Gamma Arietis, MESARTHIM, both in the "left Horn of the Ram", combine Martian and Saturnian powers, and this makes their nature into a violent one. In relevant connections, danger is indicated when acting impulsively and in a foolhardy fashion.

Take the case of a native, born 23.4.1911, Sun in conjunction with this fixed star. He went to the rescue of a friend in danger of drowning, and he himself got drowned.

In the natal chart of rapist-killer KUERTEN, Venus was in conjunction with the horn of the Ram, corresponding to an extraordinarily developed sensuality. Because of the Mars-Saturn character, such a native did not hesitate to slay.

ELSBETH EBERTIN noted, in the natal charts of a number of soldiers of the First World War, that the influence of the fixed stars showed up the fact that these people were daredevils and bold.

Mars/Saturn powers were evidenced in ROBERT OPPENHEIMER, born 22.4.1904. He had the Sun in conjunction with the fixed star. OPPENHEIMER contributed more to the development of the Atom Bomb than any other person. However he intended to use atom power for peaceful purposes. His conscience made him object to the building of the Hydrogen Bomb. His government however did this and it became a means of destruction and many people died by it.

8. EL-NATH, 6°59' Taurus, Alpha Arietis, 2nd magnitude,

main star in the "Head of The Ram", also called HAMAL, and is equivalent to an unfortunate Mars/Saturn combination. It appears as though Mars and Saturn are fighting each other for domination, and this influence could be a dangerous one in the material sphere. If connected closely otherwise, life may be repeatedly in danger, for example, if the native undertakes climbing - up or down; or by leap or dive into an abyss. If connected with beneficial stellar bodies, the influence of this fixed star will be lessened. If Mars or Venus are linked with El-Nath, and if other factors are equally present, suffering connected with love life and ill treatment by sadism is indicated.

ELSBETH EBERTIN kept an eye on the cosmogram of a fighter pilot who had Sun in conjunction with El-Nath; his rise in life was a quick one, but he was also severely injured several times.

A woman, born 27.4.1891, had in her natal chart Sun conjunct with El-Nath; she had to enter hospital because of hysteria. On May 7th, 1923 she screamed and complained about weird feelings - and then leapt from the window. She survived this fall with minor injuries, but had to enter a mental institution because she was considered obsessed. This person, though, had a further configuration in her chart, influencing her condition. Mars and Neptune were in exact conjunction, 1° further was Pluto, and this group was in conjunction with the Hyades.

ELSBETH EBERTIN had in her collection several charts of 1881, and Saturn in these was conjunct with Alpha Arietis. These people dived from heights and also got head injuries in other ways.

RUDOLF HESS, born 26.4.1894, deputy of ADOLF HITLER, got himself imprisoned by a leap from high power so to speak, brought about by conjunction of Sun with El-Nath.

The spy MATA HARI, born 7.8.1876, had Neptune conjunct with this fixed star. She died by execution on 25.7.1917.

9. SCHEDIR, 7°07' Taurus,

main star of Alpha Cassiopeia, has a Saturn nature, lessened by Venus. This fixed star incorporates serious demeanour with joi-de-vivre. Should good living transgress reasonable limits, the results will be disadavantageous.

The Arab name "Sader" means "bosom of Cassiopeia". This fixed star is not really considered unlucky, but a certain amount of demonic power seems to be ascribed to it; critical results must be ascribed to El-Nath who is only at 8' distance.

In the natal chart of the well known poetess of ballads, AGNES MIEGEL, we find Neptune connected with Schedir. Perhaps this configuration contributed to the fact that the poetess, besides having a close appreciation of home and country and of her native soil, also showed a preference for the mystical and enigmatical, the border region between the real and the un-real. The Sun in Pisces also contributed to this.

The Swedish novelist SELMA LAGERLOEF, born 20. 11. 1858: Pluto (square Mars) was in conjunction with Schedir in the personal midpoints to Sun/MC = Moon/MC.
Conjunction Moon: SCHACHT, former President of the German Federal Bank.
Conjunction Mars: Dr. RUDOLF STEINER, founder of Anthroposophy.

10. ALAMAK, 13°34' Taurus, Gamma Andromedae.

This pale blue-green star in the "Feet of Andromeda" has a Venus character and a faint Jupiter like influence. It is supposed to give a cheerful nature and a liking for change and diversion and amusement. The popularity of these people will bring benefits from others.

Alamak is a double star. It has a yellowish-reddish companion which is supposed to have a Martian character. Both stars are moving around a centre point. This is supposed to bring about the change in preponderance of the Venus and Mars character alternatively.

Examples:

Conjunction Moon: SVEN HEDIN, KARL MARX, ERNST JUENGER.
Conjunction Mars: STALIN.
Conjunction Jupiter: Graf LUCKNER, world-wide traveller and sea-farer.
Conjunction Uranus: Prof. NEISER, a Dermatologist.
Conjunction Neptune: ELSBETH EBERTIN, who suffered many set-backs and disappointments.
Conjunction Pluto: RICARDA HUCH, COURTS-MAHLER. (Each conjunction with the major planets will of course be found in charts of many people; other factors must therefore always be considered.)

11. MENKAR, 13°38' Taurus, Alpha Ceti, 1st. magnitude,

also called "Menkub", Arabic "mischir", in "the Neck of the Whale". This star has a Saturnine character, corresponding to impediments of many kinds, worries, and tests of endurance, in some kinds, hardening and toughening these particular individuals. ELSBETH EBERTIN, in March 1939, noted in the magazine "Mensch im All" (now "Kosmobiologie"):

"In some cases with conjunction of Saturn, Mars or Neptune, diseases of the throat, inflammation of larynx, sometimes death or danger by suffocation have been noted. These people are advised to take good care of themselves if they have a tendency to throat trouble, and to take prophylactic measures. They should avoid over straining their larynx and, should they have special demands regarding the larynx, they are advised to take trouble in the methodical training of their vocal chords etc."

These notations were seen to be the case only too clearly with ELSBETH EBERTIN herself. In her natal chart, we find Neptune in conjunction with Menkar on the Descendant, the house of "open enemies". All her life, she not only suffered from sore throat, diphtheria and other throat troubles, but had to fight unjustified enmities. Several times, she was nearly suffocated by gas. She survived all this, but was always afraid of death by suffocation. This fate finally overtook her when Freiburg was laid in ashes. She broke through a basement wall to the neighbouring house while the attack lasted, but she perished in the smoke.

VEHLOW (12) has noted epilepsy in several cases when Sun was conjunct with Menkar.

Please note carefully: There is a considerable difference in Right Ascension, declination and latitude between Menkar and Almanak. However, the ecliptic (longitudinal) positions of both fixed stars coincide closely. Therefore, a blending of the influence of both fixed stars is the case.

12. ZANRAK, 22°50' Taurus, Gamma in the river Eridanus,

has a Saturnian character. Anyone who has this fixed star connected with a planet in his chart should endeavour not to take life too seriously and not put too much weight on everything other people say. This person should struggle to overcome melancholy. Otherwise, this star could trigger off fear of death and suicidal tendencies.

ELSBETH EBERTIN noted several cases in which the Sun connected with Zanrak brought long drawn out illness and difficult times in life. For example, a native, born 13. 5. 1909 (Sun conjunct Zanrak) lost his life in a train accident on 13. 3. 1926.

VEHLOW (12) described the last degrees of Taurus and first degrees of Gemini as "an unfortunate corner".

In the natal chart of RAINER MARIA RILKE, Pluto is conjunct Zanrak. Of course, this conjunction is found with many people, but, in this case, Pluto is in square to Saturn in the half sum Sun/Neptune. This brings to mind that RILKE always remained solitary and could not really master life. Fear, terror and fright are words which turn up in his poems again and again, and are witness to his soul having been torn apart.

MATTHIAS WIEMAN, the actor, born 23. 6. 1902, has Venus conjunct Zanrak. He, also, was a lonely soul, full of introspection, and did not like the exposure which the limelight brought him. Yet he was tremendously successful. The moving pictures, in which he was cast as a leading man are significant: "Das Herz muß schweigen," "Melodie des Schicksals", "Die ewige Maske", "Das andere Ich".

ROBERT OPPENHEIMER, father of the atom bomb, has Mercury conjunct Zanrak. In order to follow his particular line of studies, he bought a ranch in the desert. Here, he was in complete isolation from other people. He says of himself: "There, I never read a paper, nor had a telephone. Politics and economics are not my interests. The "social" life does not interest me".

In the same way, we find Mercury conjunct Zanrak with MAX PLANCK, (born 23. 4. 1858, died 1947) the great scientist and man of learning. He was known to be of humble demeanour. When aged and having been dealt with badly by life, he lived in two small rooms as a sub-tenant, all scientific work unattainable to him, and his own scientific work having been taken away from him. Soon after

20. July 1944, his son Erwin, a conspirator (in the plot against HITLER, was hanged.

13. ALGOL, 25°28' Taurus, Beta in Perseus,

Head of Gorgo Medusa. This name is derived from Arabic "Al Ghoul" meaning "demon", "evil spirit" or "devil". Derived from the same root is "Golem" of Prague and "alcohol". Algol is part of a double star system. Its darker brother circles the brighter star in about 69 hours, in such a manner that an occultation for 9 hours appears as viewed from Earth, and this gives a periodical change in brightness. Continually, its brightness alternates between values 2,2 and 3,5. It is assumed that other dark stars belong to the solar system of Algol.

The dark greater one has the property of pure Saturn character; the lighter one corresponds not only to Saturnian influence, but also to Mars-Uranus-Pluto nature. If the dark fellow is showing towards earth's path, the destructive invisible action is exercised. These will be the hours in which Algol is least bright. In the olden times, people were well aware of this.

Dr. LOMER wrote in "Kosmobiologie", August 1950, Page 302: "Arabic commanders in chief, in times of conquest, made it a point that no important battles were begun when the light of Algol was weak."

In spite of the appreciable distance from the ecliptic, this double star's influence is strongly felt and in most cases it is a disastrous one. Primitive natives will be of the base mind with an inclination to brutality and violence, especially when Algol is found together with Mars or Saturn.

In the cosmogram of mass-murderer HAARMANN, Algol is positioned on the Ascendant and in conjunction with Mars. This configuration pointed, in his case, not only to his own murderous nature, but also to his own execution.

Mass-murderer KUERTEN also had Algol with the Ascendant and in Mars/Saturn midpoint. In addition to that, he had Elscheratain conjunct Venus, and this, with his other configurations, is in accordance with the chart of a rapist-murderer.

According to ASBOGA, (13) it is very difficult to avoid accidents and severe injuries if Algol is in conjunction with Sun, Moon or the "Malefics".

ELSBETH EBERTIN's cosmogram shows Sun conjunct Algol and only 2 1/2° distant from Pluto. She could not possibly harm even an animal. But, in spite of her knowledge that she was constantly in danger, she could not avoid the air raid on Freiburg, though she was warned shortly before.

Mars in conjunction with Algol was in the chart of BORMANN, one of HITLER's closest colleagues, who had many murders to his name. We find Algol conjunct Pluto in the cosmograms of Dr. LOMER, writer who specialises in borderline science, and H. NUESSLEIN, the medium-painter, and STALIN the dictator.

Mars was in conjunction with Algol when MARCONI was born. He pioneered wireless telegraphy, and he had a tragic end.

Neptune in conjunction with Algol was significant in Admiral CANARIS, chief of German counter-espionage.

This short survey could be extended without much effort. As everything has two sides, it has to be said that "high spiritual rays" are emanating from Algol also, but only those human beings can receive them, who have already reached high spiritual development. Even then, it has to be noted, that those particular persons will have difficulties and obstacles in their way, and they have to use much energy to overcome these handicaps. If their endeavours fail, strong counter forces and enmities will be present.

In the journal "Kosmobiologie", 19th. year, 1953/9, another confirmation of Algol is recorded:

The fixed star Algol is, according to the 1950 ephemeris, in 25°28' Taurus. One of our readers noted that, in connection with Algol, artificial teeth are often the case. Again and again it is confirmed that life contains great handicaps (or good fortune is rare) if Saturn and Mars are found on the same place.

The following examples are given:

1. Female, born 29. 5. 1912, Saturn 25°25' Taurus. Her husband was

accounted among the missing personnel in the battle of Stalingrad, 1945. She fled via the Baltic Sea, was raped by Russians in 1946, contracted syphilis which was not detected until too late, lost all her teeth: had to put up with dentures, lost much of her hair, had a weakening heart condition. She married again 21.11.1952.

2. Female, born 28.7.1881, Mars 25°43' Taurus: Had dentures.
3. Male, born 23.3.1940, Mars 24°08' Taurus: Lost incisor, just as though it was cut off, when engaged in diving exercises.
4. Male, born 2.8.1885, Neptune 25° Taurus: Had his own teeth until his death occured in 1948. His teeth however were very poor ones. He also had a hard fate.
5. Female, born 14.8.1885, Neptune 25°32' Taurus: Artificial teeth, in old age many illnesses, especially rheumatism.
7. Male, born 31.1.1887, Neptune 25°02' Taurus: Very poor teeth, mentally handicapped, died end of June 1948.
8. Male, born 4.8.1910, 4.30 am Allenstein/East Prussia, Lunar Node 24°30' Taurus: Died with his comrades from German bombs as a P.O.W. in Jugoslavia (Node = in a community or association).

It would be desirable if readers of this book would work and record similar line-ups of the diverse fixed stars. The more examples are available, the easier it is to recognize the meeting points between fixed stars and happenings in life.

14. ALCYONE, 26°19' Taurus, eta tauri,

was called after the Greek nymph, and is main star of the Plejades. These stars are called in the vernacular "the seven fixed stars" or the "clucking hen". The "Plejades" in Greek means "the pigeons". Nine stars of this cluster of stars have their own names. Six or seven can be seen with the bare eye. The stellar cluster appears as a nebula, because another 200 or so telescopic stars belong to it extending over several degrees. Therefore Alcyone as representative of all Plejades is given a larger orb than is usual.

ANNA CREBO, Editor of the Journal of Astrological Studies, USA, put in "KOSMOS" (14) the centre of the Plejades as 29° Taurus.

The Plejades correspond in their influence to a combination of Moon and Mars. If well connected otherwise, this points to ambition and endeavour, which have as their result preferment, honour and glory.

The power of Mars is good in raising people to high positions. However, with other relevant planetary connections, this can make for high passion and ruthlessness. The best example here is STALIN. Alcyone, Pluto and Algol are in conjunction in his nativity. Unfavourable connections bring the danger of enmity and fall from power.

The influence of the Plejades has sometimes been noted as not a good one for eyesight. Disturbance of proper sight, injuries to the eyes, even blindness (especially when Sun and Moon, the lights, are aspected by it closely) and when the Plejades are in critical aspect to Mars or Saturn. As an example here, the blind English poet MILTON should be noted.

ELSBETH EBERTIN collected a great many examples of a connection of the Plejades with diseases of the eyes and blindness. Her records however were lost in the fire of Freiburg. If someone detects a link with the Plejades in his chart, this person should always be most anxious to preserve his eyesight.

To continue with the influence of the Plejades ELSBETH EBERTIN points to the fate of the Czar's family in her "Blaettern zur Einfuehrung in die Wissenschaft der Sterne" (Introductory notes into the science of the stars) (15) and 1915 in her book "Koenigliche Nativitaeten" (Royal Nativities) (16). Czar NICOLAUS II had a conjunction of Sun and Mars with these seven stars. ELSBETH EBERTIN remarked here:

"Sun conjunct with the Plejades in most cases brings a violent death by blow, stabbing or murder."

This prediction was borne out on 17. 7. 1918 when the Czarist family were murdered in a most horrible manner.

Natives born with Sun positioned in the Plejades, however, can reach old age, and achieve fame and honour. The cosmogram of the well-known painter HANS VON VOLKMANN, born 19. 5. 1860, shows Sun conjunct Plejades. This man passed away following a sore throat after an operation. Therefore, configurations pointing to violent death caused by another person can show in death as a result of an operation.

Lastly, it has to be mentioned that Alcyone is not a good omen with regard to relationships to the opposite sex. Connected with Neptune, it is supposed to give inclination to homosexuality.

HILDEGARD KNEF has sometimes been thought of as "a sinner". When she appeared in a movie without clothes, she caused a sensation and she was criticized severely. The movie-actress has, in her nativity, the Plejades exactly on her Descendant.

15. HYADES, 4 - 6° Gemini,

cluster of the rain stars, or the "Regen-Gestirn" - rain drops, is still in the sign of the Bull. These stars bear out a Mars/Neptune character with an Uranian blend. The influence is that of a staggering increase of the sexual urge, self preservation which could lead to greediness, sexuality, a dissolute life, excessive and licentious ways, and a striving for prestige leading to power politics. If this tendency is bridled, an increase in energy in the positive sense can be the result, leading to tremendous success in life. When the pinnacle of the career is reached, the danger appears that the inherent nature may come again to the fore, and may, if uncontrolled, lead to failure and to a fall from power. With a connection to the Hyades, it is doubly important that power and strength must not be used to exploit other human beings. VEHLOW (12) recounts that "The Hyades with Neptune in the 8th. house cause confused ideas, making "world saviours", and leading to delusions of grandeur". VEHLOW did write this before 1934. It is therefore hardly possible that he meant this to refer to HITLER in whose nativity Pluto is found in conjunction with the Hyades and nearby positioned is Neptune in the 8th. house.

The Hyades on the MC or AS and connected with Sun or Moon is supposed to give the aptitude for a military career. Connected with Saturn, the Hyades point to falls and accidents.

CZAR NICOLAUS OF RUSSIA,

born 18. May 1868 about 12. 02 p. m. in Petersburg.

Abdication 15. 3. 1917

Assassination 17. 7. 1918

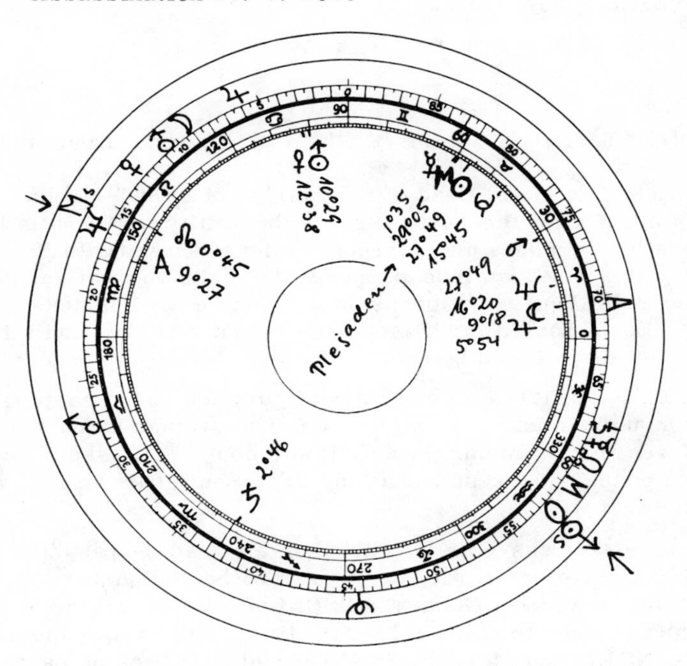

The figure above shows how the Sun forms a conjunction with Alkyone and MC in the centre of the Plejades. When the Czar abdicated, the beginning of a time of suffering is indicated, followed by his assassination. At this time Uranus advanced by measure of Solar Arc, came near to making a square to the Sun, and this is especially easily seen in the outer 90° sphere. Meridian s made a square to Neptune which itself is in semisquare to Saturn and can be seen in the 90° sphere directly opposite of it. The radical Uranus had gone over the square to the Sun when the Czar was murdered, in such a manner that Uranus s acted as direction and transiting Uranus triggered off this event in a life thus fated.

Sun conjunct Hyades is to be found in the cosmogram of TITO. Pluto in conjunction with the Hyades is in Chiang Kai Chek's cosmogram and also in EISENHOWER's. Graf LUCKNER, well known captain of the "Seeteufel" (= German for "Devil of the Seas"), has his AS with the Hyades. These examples show the conjunction of the Hyades borne out in a positive way.

16. NORTHERN BULL EYE, 7°27' Gemini, Epsilon Tauri,

the "Right Eye in the Head of the Bull" is of a Venus-Moon character. Linked with the Sun, it gives the ability to get on in life, to achieve high honours and gain possessions. Linked with Neptune or Venus, this fixed star is supposed to make the native artistic. Linked with Jupiter, success in art or the art business is indicated. But if Saturn is unfavourably linked with Sun and this fixed star, rise is followed by fall.

ELSBETH EBERTIN noted this configuration in the cosmogram of a distinguished leader of commerce. This woman had accumulated great wealth but, during the inflation (about 1923), she became almost a pauper as a result of a long drawn out court action over her divorce.

The movie-actress GRETA GARBO, also styled as "the divine GARBO", made her career as a result of the conjunction of Jupiter with this fixed star. General EISENHOWER had Pluto with this star and he used to paint in his free time. The French movie actor FERNANDEL has Mercury 7°34' Gemini. He became particularly well known by his taking the part of the priest in the film DON CAMILLO. ELLY BEINHORN, aviatrix and gifted writer, has Sun in 7°55' Gemini. EDISON, the successful inventor, has Jupiter in 6°50' Gemini.

17. ALDEBARAN, 9°05' Gemini, Alpha Tauri,

the "right and bright eye of the Bull", called Lampaurus by PTO-

LEMY, the beacon, Royal fixed star of Mars nature, called Aldebaran by the Arabs, meaning "the one who follows the Plejades". If conjunct with AS or Sun, this will point to extraordinary energy. Such a person will be able to be ahead of other people, will gain a leading position, and will be acknowledged, but also will make enemies, through whom dangers will threaten him.

Saturn in conjunction with Aldebaran and linked in an unfavourable manner with other stellar bodies could bring danger and loss through water, e.g. by floods, storm, shipwreck, drowning.

Uranus in conjunction with Aldebaran gives extraordinary energy and capacity for work, whereby the native gains prestige and appreciation. On the other hand, the opposition of opponents is evoked by it. Dr. RUDOLF STEINER, the successful but also highly controversial founder of Anthroposophy, has this configuration in his cosmogram. Also the former King FERDINAND of Bulgaria had in his cosmogram Uranus conjunct Aldebaran. This man did play an important part in politics in his time. However, he also had a great many opponents.

In world events, the conjunction of Mars or Saturn with this fixed star corresponds to catastrophes caused by weather conditions - e.g. floods, shipwrecks.

According to tradition, the Moon with Aldebaran points to danger through poison, if Neptune is also lined up in this configuration.

Venus connected with Aldebaran is analogeous to "power directed wrongly" and abnormalities in love life. This statement has to be seen in context, however.

Aldebaran is almost exactly opposite to Antares which is 9° Sagittarius. In USA, both fixed stars are described almost as a unit. Frau Dr. INGE KOCH-EGENOLF (17), born 12.8.1910, has Antares conjunct MC and Aldebaran conjunct IC. She ascribes to this her chronic sclerotic illness, luck and bad luck through domestic animals, as well as a psychological clairvoyance.

REINHOLD EBERTIN in "Pluto-Entsprechungen" (18) cites in figure 48 the cosmogram of an "abnormal man" with Mercury in 9°34' and MC in 9°00' Gemini and opposition to Saturn in 9°50' Sagittarius and Venus 45° Pluto. This is a case of a malformation of the urinary tracts and the sexual organs. After several operations, the danger of sterility is indicated.

In the same book, a further abnormality is recorded in figure 107. Sun is 8°44', Pluto 9°14' and Neptune 10°50' Gemini in conjunction with Aldebaran and corresponding in this case to a congenital paralysis. In addition to this, the left arm is missing altogether, the right one is stunted and hanging down limply, and the legs are not properly developed and cannot be used. This man has to rely on his teeth. He paints with brush in mouth and earns a living on the fair-ground.

In the same book is recorded a case of prostrate trouble, figure 108: Pluto 8°29', Neptune 9°09' Gemini in semisquare to Venus.

The heavy planets, e.g. Neptune and Pluto, can only be used in examples, if other aspects also contribute, as the basic pattern of these planets is present in many natal charts.

18. RIGEL, 16°08' Gemini, Beta Orionis,

on the "left Knee", Arabic "foot, support", in the constellation of Orion, corresponds to a Mars-Jupiter combination. If Rigel is conjunct with Sun, Moon, Meridian or Ascendant, quick rise in life is promised on account of a strong inherent will power, love of action, and a lucky hand in enterprise. A continous battle to retain an acquired position has to be waged. This fight at the same time stimulates an increase in vigour. Even in spite of unhelpful aspects, success and reaching the set aim can be secured by this powerful concentration. If however the native is not cautious or has an attack of weakness, failure and disappointments, a fall from success will follow.

Tradition regarding the influence of Rigel is quite contradictory, implying again that the cosmogram must be fully analysed.

After the First World War, at the time of the first trans-atlantic flights, ELSBETH EBERTIN noted a connection of Rigel to Sun and Moon, in the charts of several world record pilots, leading to sudden difficulties in carrying out of planned enterprises resulting in failure to reach the set target.

With this Mars-Jupiter nature, Rigel has particular meaning when found in charts of government officials, military men, politicians, leaders of political parties, barristers and priests.

Examples:
Rigel conjunct Ascendant is found in the cosmogram of the first President of the German Republic, FRIEDRICH EBERT; in the cosmogram of the politician and minister ERZBERGER, murdered 1921.
Rigel conjunct Sun: poet THOMAS MANN.
Rigel conjunct Moon: first trans-atlantic flyer Captain KOEHL, and also General FRANCO.
Rigel conjunct Venus is to be found in the nativities of MAX VON SCHILLING and President J.F. KENNEDY.
Rigel conjunct Mars is significant for Graf ZEPPELIN, who first had failures, but did not give up his endeavours with his air-ship, and writer ERNST JUENGER, whose writings speak for a fighting spirit.
Rigel conjunct Pluto = Mercury / Uranus is in the natal chart of REINHOLD EBERTIN.
Rigel conjunct Neptune could have been decisive in German Federal President Dr. LUEBKE not being properly appreciated. In the cosmogram of General DE GAULLE, Rigel is conjunct Lunar Node and square Saturn.

19. BELLATRIX, 20°16' Gemini, Gamma Orionis,

"on the left Shoulder" of Orion, is pointing to its character by its name "The Amazon", being equivalent to Mars action. In addition to this, there is an influence of a Mercurian kind. This will make for characteristics such as quick decision taking, thoughts and plans being realized with energy, courage, fighting spirit, strategic talents, ability to organize, discrimination, - but also often the reckless aggressiveness of a belligerent dare-devil. If the positive properties can be drawn out, connections of Rigel, e.g. with

MC, will lead to advancement and success. But those who succeed always have to allow for being surrounded with envy and hatred.

Bound up with the Moon, dangerous and severe injuries have several times been recorded, including eye injuries. A nurse who had Pluto and Neptune in conjunction with Bellatrix in her nativity, became blind suddenly after treatment with ultra violet rays, (not applied properly, we assume). A native, born 27.1.1898, was called up during the war and put into battle lines and on 23.2.1916, he received injuries by a shell splinter in his right eye. He got further shell splinters on 18.11.1917. He had Neptune conjunct Bellatrix.

An entirely different match was evident in a native, born 21.2.1898 with conjunction of Bellatrix and Neptune. He was indicted for high treason. The indictment was eventually withdrawn but the native had to go through dreadful excitement and inconvenience.

The stigmata THERESE NEUMANN was born when Neptune was conjunct Rigel. She was blind for four years. It appears that MARLENE DIETRICH, a movie-actress known all over the world, has been helped in her career with a position of Bellatrix conjunct MC. The American businessman N. A. ROCKEFELLER has a Moon Bellatrix conjunction. Dr. GOERDELER, chief of the underground working against HITLER, was executed; he had Saturn in conjunction with Bellatrix.

20. CAPELLA, 21°10' Gemini, Alpha Aurigae,

main star in constellation Auriga (the charioteer). Alpha Aurigae means "little goat" or "goat star" called so because "the goat" carries "the charioteer" on her left shoulder. Capella also has a Mercury-Mars nature. The Mercury properties are more eminent, and show in love of learning, studiousness and interest in research. These properties are accentuated if Capella is connected with Moon, Mercury or the Ascendant. In plain people, these properties make themselves known by persistent, annoying and inquisitive curiosity. According to tradition, this fixed star also makes people somewhat odd, and there is a tendency for such natives to cut capers. This could be counted as another manifestation of a weak Neptunian influence. Connected to beneficial stellar bodies, natives will become popular, receive honours and have success in material enterprises.

Capella in conjunction with Jupiter, is found with Prof. KARL JASPERS, co-founder of existentialism and also with the novelist JOS. AUGUST LUX. When WERNER HEISENBERG, the scientific revolutionary, was born, Capella was on the Ascendant, as was also the case with the poet CHRISTIAN MORGENSTERN who, in addition, has the company of Venus there. Capella in conjunction with the Moon is in accordance to the nature of the actor HANS SÖEHNKER. The influence exerted by the former President of the German Bundestag, HERMANN EHLERS, tallies with Pluto in conjunction with the fixed star and Mercury square to it.

21. PHACT (Noah's pigeon), 21°29' Gemini, Alpha Columbae,

corresponds to the nature of Mercury-Venus, with a touch of Uranus. This fixed star in a good conjunction is supposed to give an appreciation for form and rhythmics as well as artistic talents, and is also supposed to confer ardent interest in science. Phact near the Sun is found in the cosmograms of the great mathematician NEWTON and the composer RICHARD STRAUSS. This fixed star gives a touch of genius and medium-ship. It is assumed that this fixed star will gain importance with the beginning of the "Aquarian Age".

22. POLE STAR, 27°54' Gemini, Alpha Ursae minoris,

main star of "the small Bear", situated on the tail, has a Saturn nature, combined with qualities of Sun and Venus. It might seem strange to include this fixed star here, as its latitude is about 60° and therefore placed far outside of the ecliptic in which the planets move. Measured on the ecliptic, it is situated closely conjunct with Alpha Orionis, Betelgeuze, the main star of Orion.

The Chinese considered the Pole Star as "the great honourable Lord of the Heavens". Other races too had high admiration for it, and one can draw the conclusion that, in a relevant position in the natal chart, it will give spiritual powers to the bearer, and he will be highly respected.

The Pole Star serves as a guide and indicator. If it is conjunct with planets in the angles, the native will have a good sense of discretion and is able to follow "his instinct". He clearly recognizes his aims, and will pursue and achieve them.

In checking these remarks with the records, one will find a great deal of confirmation. IGOR SIKORSKY the Russian aeroplane builder, has Mercury in conjunction with the Pole Star, and he has a reputation for "mastering the impossible". WERNHER VON BRAUN, designing engineer of the flying bomb, has a conjunction with the Pole Star and Ascendant and Pluto = Sun/Mars. The physician and psycho-therapist FELIX SCHOTTLAENDER has his Sun here, but SIGMUND FREUD, founder of modern psycho-analysis has his Saturn here. This will explain why "this great man was poisoned by not being recognized. Even when aged 53 years, he was still treated as an outcast in Europe".

The well known air ship skipper Dr. HUGO ECKENER, had a Mars conjunction with the Pole Star. The novelist RICARDA HUCH, called "the most refined and best talented of the last century", had on this degree of the Zodiac Uranus = Mars/Jupiter, and LILLI PALMER, the film actress, had Venus connected with this fixed star. The English novelist, CHARLES DICKENS, and the German poet CHRISTIAN MORGENSTERN have Jupiter in conjunction with the Pole Star. The negative outcome is shown in the conjunction with Neptune which is shown in the chart of the deposed dictator of Indonesia, SUKARNO. Also the same Neptune conjunction is found with O. E. HENSCHEL, owner of a large industrial plant in Kassel (manufacturing railway engines, trucks etc.). He had to give up his enterprise (Neptune = Sun/Mars).

23. BETELGEUZE, 28°04' Gemini, Alpha Orionis,

on the "right Shoulder" of Orion, and by its red colour complying to the meaning of Mars with a touch of Mercury. It has been observed that the effect of these properties is more propitious than those of Bellatrix, especially if Jupiter is connected with it also. Placed in the angles, preferment, luck, success and even everlasting fame are promised. On the Ascendant and in conjunction with Mars there will be danger of death by lightning, fire, explosion, fire arms or similar things. ELSBETH EBERTIN recorded in June and July 1927, when the rising Lunar Node was in conjunction with this fixed star, that groups of people died by lightning or were victims of explosions or assassination.

As one can see by looking at relevant positions, Betelgeuze and Pole Star are concurrent with only 10' longitudinal distance from each other. Their latitudes are very greatly different however. Betelgeuze is situated in approx. 16° Southern latitude and the Pole Star in approx. 66° Northern latitude.

24. MENKALINAM, 29°13' Gemini, Beta Aurigae,

in "the Head of the Charioteer", is basically of Jupiter character, with a blend of a weak influence of Mars and Venus character. Only when really in good aspect is this fixed star rated as a positive acting one. Badly positioned, this star will be most unhelpful. If Jupiter is found in the same degree as Menkalinam, the native has a promise of fortunes, honours, riches, popularity and exceptional friendships. In conjunction with Mars, it has been found that death has occured on the occasion of official festivities, military exercises, or battles.

The fabulously wealthy "diamond king" HARRY OPPENHEIMER has Lunar Node in conjunction with Menkalinam. The writer, known for his writings on sexual enlightenment, OSWALT KOLLE, was exceptionally successful, and in his cosmogram Mars is positioned conjunct with this fixed star.

25. ALHENA, 8°24' Cancer, Gamma Geminorum, 2nd magnitude,

situated on the base line of constellation Gemini, is supposed to have a Venus nature with Jupiterian influence. This will confer a spiritual orientation on people thus affected and give also artistic inclination with interest in the sciences.

Significantly, the composer ROBERT SCHUMANN had Venus in conjunction with Alhena. The spiritual influence is shown in the map of the medium MARIA SILBERT who had Uranus = Sun/Neptune in this degree of the Zodiac. The mountaineer HILLARY was compelled by thirst for knowledge and love of nature to climb the highest peak on earth and to research the South Pole. It is significant that Mars, the planet of energy, was in conjunction with Alhena at his birth. The first German opinion pollster, NOELLE-NEUMANN, had in her natal chart Ascendant = Mercury/Pluto and this complex connected with the fixed star.

26. SIRIUS, 13°23' Cancer, Alpha Canis maioris,

"in the Neck of the Great Dog", "the shining one", "the lustrous sparkling bright one" is the brightest star in the Northern hemisphere. In mythology and conjuring magic, it played an important role. There were Sirius sanctuaries in Egypt, Persia, Greece and in Rome. The old Germans called it "Lokis Brand" (= fire of Lokis). This star is "only" 8 light years distant from us. In its nature are Mars and Jupiter traits. From time immemorial it was the "Royal one", but it is also known as violent. Well connected, it promises fame, honours, riches. On the Ascendant and with Mars combined, Sirius can be quite dangerous. Pushing ahead with too much ambition is then seen, resulting in dangers by injuries or attempts on the native's life. According to tradition, Sirius will give a "famous" death with honours beyond the grave, if positioned in the 8th house.

In good aspect with Mars and Jupiter and close to the Meridian, promise is given of gaining extensive wealth, a lucky hand in commercial enterprise or matters of government. This star so placed is most excellent for the military, lawyers and civil servants.

Sirius in conjunction with Sun and well placed in the cosmogram will be found to be the case with numerous important and famous personalities. Possibly, the rise in station is made possible by protection from people of influence.

Sun conjunct with Sirius was the case with President of the USA, COOLIDGE, and the poets PAUL KELLER and HERMANN HESSE. HJALMAR SCHACHT has his MC with this star. He was called the German "financial wizard".

Prof. ALBERT EINSTEIN has Sirius = Ascendant = Jupiter = Uranus/Pluto. H.P. BLAVATZKI, founder of the theosophical movement, has Sirius = Ascendant = Pluto = Mercury/Jupiter.

WILLIAM BOOTH, founder of the Salvation Army has his Moon positioned with Sirius.

The stigmatized Pater PIUS has Venus conjunct Sirius.

We find with the discoverer of radium, Prof. Dr. MARIE CURIE, that Sirius links up with Uranus = Mars/Jupiter. Prof. HANS DRIESCH, the important biologist, philosopher and parapsychologist, has Sirius = Uranus = Neptune = Mercury/Jupiter.

27. CANOPUS, 14°16' Cancer, Alpha in "the Boat Argo",

was given this name, because the ancient ones thought they saw a ship in this particular part of the heavens, and they believed it to be a wise step to relate it to their patron Canopus, the old Egyptian god, who was the patron of skippers and voyagers. Combined in Canopus are powers ascribed to Jupiter and to Saturn.

ELSBETH EBERTIN has recorded several cases of persons, born early July and in the morning between 5.00 and 6.00 a.m., who made extensive sea voyages.

Linked with a badly placed Saturn, the tendency to depression and even suicide is inherent. However, if Saturn and Jupiter are well placed in the chart, harmonising and spiritual powers are evoked. According to tradition, Canopus on the Ascendant is supposed to give a love of travel, and also to instigate fights and quarrels, resulting in law suits: the native however can channel these tendencies constructively, with astuteness and a sense of real earnestness.

Links to the fixed star Canopus have been noted several times in the nativities of writers and film actors, and especially of those persons who, in our times, have to undertake many journeys in connection with their position:

Film Star EMIL JANNINGS has Canopus in conjunction with Venus, and so also has LEO SLEZAK. MARIA SCHELL has Canopus conjunct with Pluto and opposite the Sun, the writer H. PIONTEK has Canopus conjunct Pluto = Venus/Jupiter. MATA HARI, the spy, had in her nativity this fixed star in conjunction with Venus = Jupiter/Saturn = Saturn/MC. In INGRID BERGMANN's map, unpleasant experiences in her marriages are indicated rather in the conjunction of Saturn = Moon/Mars.

On the other hand, Canopus conjunct Sun is found with the communist woman KLARA ZETKIN, conjunct Moon with the poet HERMANN HESSE, and conjunct Mercury with the psychologist Prof. H. C. JUNG.

28. CASTOR (Apollo), 19°33' Cancer, Alpha Geminorum, 2nd magnitude,

in the twins. and together with Pollux is called "Thiassis eyes", a name given to these two stars by the ancient Germans. The Babylonians considered both stars as a pair belonging together, as "herdsman and warrior". The Phoenicians recognized, in the twin stars, helpers if in peril on voyages at sea.

Castor is influenced by Mercury and has a blend of Jupiter in it. Linked with the Moon and Mercury, it has the effect of such people being blessed with a good nature and fine morals. It is also supposed to convey refined manners. A conjunction with Sun or Mars will make for energetic characteristics and a certain tendency for satire and cynicism, depending on the position of Mercury and Mars in the chart as a whole.

Castor in conjunction with Mars is found with ELSBETH EBERTIN who put lots of energy into defending Astrology. She conferred with scientists, newspaper editors and civil servants, battled with them and won and then could not restrain herself from making satirical remarks. Significantly, her colleague GEORG HOFFMANN has Mercury positioned in the same place in his cosmogram. The poet ERICH KAESTNER has a Castor conjunction with Mars. He is well known for his biting criticism of society and he put before society his "distorting mirror" ("Zerrspiegel" in German - a journal). Also known as a critic is HANS MAGNUS ENZENSBERGER, who created and edited poems which criticized the time and, hate-criticising poems. He had a name amongst his contemporaries as "the angry young man". In his nativity, Castor is in conjunction with Pluto = Mars/Jupiter. The Beethoven pianist ELLY NEY always reserved her own opinions. Castor was conjunct her MC at her birth. THOMAS MANN had a conjunction Canopus-Moon. A tragic life was the destiny of MARIA ANTOINETTE, Queen of France, whose life was ended by the guillotine. At her birth, Castor was positioned together with Mars.

29. POLLUX, 22°35' Cancer, Beta Geminorum, 1st magnitude,

also called Hercules, in the twins, has a strongly felt Martian nature and has the name "the wicked boy" of the brother-and-sister part of Castor and Pollux. According to its nature, this star is brutal, tyrannical, violent and cruel if in conjunction with the Lights, Ascendant or Meridian or even linked with malefics. Just as Mars has its good sides if the energy it creates is channelled constructively, so Pollux should not always be considered as unhelpful. The fixed star badly placed could manifest that way, not by the native misusing his energy but by others deceiving him and by fate playing him some cruel tricks.

The famous scientist Prof. MAX PLANCK had a Saturn-Pollux conjunction. In spite of having been known world wide and having had the Nobel Prize conferred to him, he had a very sad old age. His son was executed by hanging for his involvement with the plot on 20.7.1944. His home was destroyed, - and after the war, with all his scientific tools taken away from him, he and his wife had to spend the last years of his life in two small rooms as sub-tenants.

BURKHARD HEIM, the physicist, has in his cosmogram Pollux in conjunction with MC = Mars/Pluto. While experimenting, he had an accident and lost his eyesight, but he did continue to work, finding entirely new sources of energy, - attesting to both the positive and the negative sides of Pollux.

The Indian leader Mahatma GANDHI was a revolutionary in spite of his non-violence directed against the British rule, - Uranus in conjunction with Pollux is significant in his map.

Pollux in conjunction with the Sun is found with the lively SPD speaker of discussion fame in the German Bundestag, FRITZ ERLER. LENI RIEFENSTAHL had manifested also the positive as well as the negative tendency of Pollux, in conjunction with her Mars and opposite Saturn. She was the film actress and stage director who in her time made a movie of the Olympic Games in Berlin. She also made the propaganda-movie of the German Party Conventions, named "The Triumph of Will Power". Her membership in the Nazi party was, later on, a disadvantage to her.

The "violent boy" of communism, STALIN, was born with Pollux in culmination, i.e. conjunct MC.

A completely different manifestation of Pollux influence is found in prostitute, ROSEMARIE NITRIBITT. At her birth, Pollux was in conjunction with Pluto and in opposition to Venus. She became a victim of her "sex-job". She was murdered.

Once again, one has to point out that the fixed star conjunction alone is not necessarily responsible. One has to study the structural elements in the entire cosmogram. Configurations such as this conjunction would indicate the need to be careful, but one cannot deduct a violent death as the outcome.

ELSBETH EBERTIN concluded that in some years the conjunction of Saturn and Pollux works out rather in gall bladder and stomach operations, analogous to an injury to the body.

Pollux linked with Mars and Moon, in good aspects, in the experience of Frau EBERTIN, gives an inclination to travel to far distant places, or even to emigrate.

30. PROCYON, 25°10' Cancer, Alpha Canis minoris,

in the "Neck of the Small Dog", the "Dog in the front", the "Preceding Dog", has a Mars/Mercury nature, and therefore makes people hasty, jealous, pig-headed; but a later it also confers will power and ability to put thoughts and plans into action. According to tradition, there is also a tendency to hot temper and impudence. Rise and success are found with it, but fall from high position later is indicated. Enterprises created in haste therefore do not last. People who "want to go through the wall with their head" only cause injury to themselves. Procyon gives drive and a good sharp mind. Linked with positive stellar bodies, success is made greater, but the native, in order to avoid a fiasco, has always to take care not to be imprudent. Especially dangerous is Procyon configured with Mars and Pluto. If Sun is with the fixed star, a valiant demenaour is indicated.

Dr. ECKENER, the air-ship pioneer, who had in his natal chart Procyon in conjunction with MC, appeared to be aware of the inherent dangers. He never forgot "the pricks of conscience before a decision", but he did experience the Zeppelin catastrophe, at a time when Pluto s was in conjunction with MC and Procyon.

Far more critical an outcome was the link with MC in JOHN F. KENNEDY. He had MC = Mars/Pluto. He was assassinated when a direction of Saturn was due with this particular point.

The South African Premier VERWOERD had the fixed star in conjunction with the Moon. He survived one attempt on his life, but the second attempt was fatal.

ELSBETH EBERTIN as well had Procyon in conjunction with the Moon. Several times, she was the target of extreme aggression, and she did succumb in the air raid on Freiburg/Br. in November 1944.

Less disadvantageous was the conjunction of Procyon with Moon in the map of Field Marshall HINDENBURG. Perhaps one can bring a connection here with the battles of annihilation which he fought.

31. PRAESEPE, 6°34' Leo,

the "manger", a cluster of stars in Cancer, designates a part of the Zodiac which, since ages past, has had a bad omen. Because the Aselli are so near, a total of 5 to 8 degrees in the sign Leo are hit by it. The Praesepe appears as a thin stellar nebula and, in the telescope, this is to be seen as an open cluster of vast expanse. There are about 500 stars of 6th to 17th magnitude with a diameter of 13 light years. The astral influence of these far distant worlds is equivalent to a combination of a Moon-Mars emanation. Also acting is a Neptunian influence, both in Praesepe as well as in the Aselli.

Not good for head or eyes is the distinction given to Praesepe on the Ascendant. If the Moon is poorly placed, an increased susceptibility to infectious diseases is present. One can observe a craving for and abuse of stimulants, luxury foodstuffs and narcotic drugs; e.g. heavy smoking. With a really afflicted Moon, great danger to eyesight is indicated. If other stellar bodies are linked up here, the result can be quite critical.

GEORG HOFFMANN has on record a relative, born 9. 8. 1935: congenital short sight. There is no hereditary pattern of short sightedness. Here we find Praesepe in exact square to Uranus and Mars, who themselves are in opposition. The picture therefore is: Praesepe = Mars/Uranus as well as = Sun/Pluto. This weakness of eyes was alleviated by Saturn situated in 9° Pisces on the Ascendant.

ELSBETH EBERTIN pointed out that countless soldiers who became blind during the First World War, actually had in their cosmograms relevant connections to this fixed star. (These reports unfortunately were lost during the bombardment of Freiburg.)

In these days, ophthalmologists have enormous skills and many a case will not be as serious in its outcome, - therefore a conjunction with Praesepe should not be feared so much.

According to VEHLOW, the Chinese gave this group of stars the name "the spirits of the ancestors" and were of the opinion that, mainly if persons had a conjunction with the Moon, they would have peculiar experiences with the realm of the dead. The same can be said for spiritualist seances. The Neptunian impact is shown thereby.

ROSEMARIE NITRIBITT
Frankfurt "Society Lady",

born, 1.2.1933, 4.45 a.m.
was murdered 1.11.1957

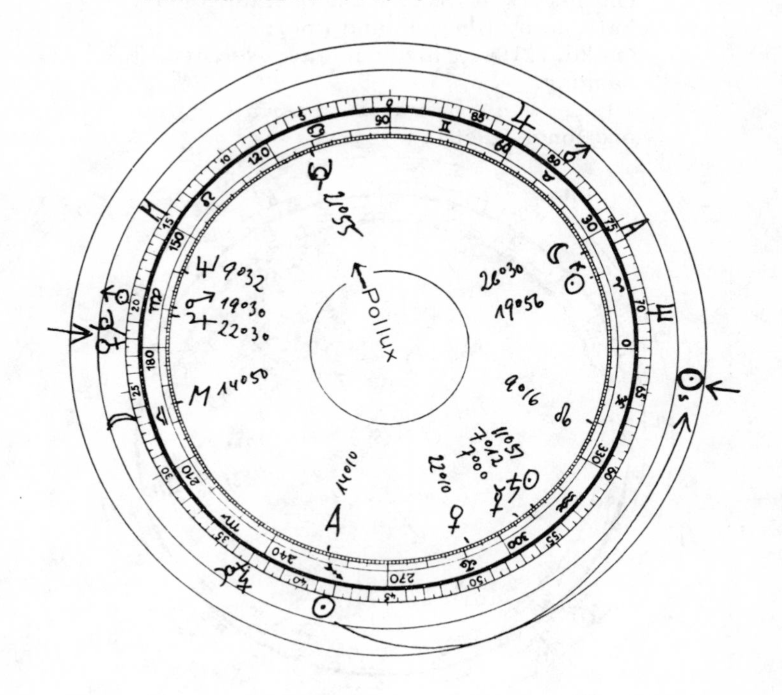

This is a characteristic example for Pollux in conjunction with Pluto and at the same time in opposition to Venus. This fixed star increases considerably the attraction and sensuality indicated by Pluto-Venus. This configuration at the same time is in the half sums of Moon/Uranus and Moon/Meridian. When Sun s picked up this connection and when the native, so to speak, was at the highest point of her "love activity", she was murdered. A marked indication for the death was MC s in square to Saturn.

On the opposite page, attention is drawn to MAX PLANCK who had a different planetary connection and also quite a different life.

Professor Dr. MAX PLANCK

born 23.4.1858 in Kiel, Germany

Discovered the law called Planck's Law. On this law is founded the quantum theory built up by Einstein and Bohr.
On 20.7.1944, his son was executed by hanging.
1947 saw his own death after a very sad and lonely last few years.

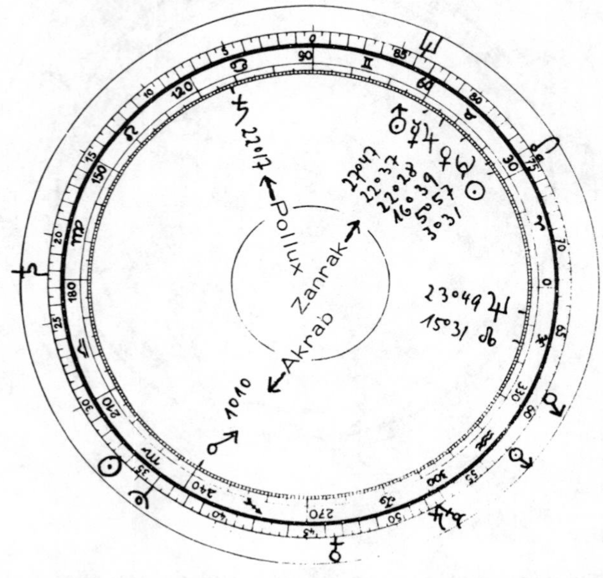

The cosmogram of PLANCK is indicative of a fated life on account of the conjunction of Zanrak with Mercury and Jupiter; his life was first dominated by extensive scientific work and the conjunction of Pollux with Saturn was discharged by heavy blows to his life when he was quite aged. An Akrab-Mars conjunction is a significant pointer to mass catastrophes. One which hit him very hard was the legal execution of his son who was a member of one of the underground movements against Hitler. World War II was a mass catastrophe which cost him all his belongings, his home and book collection.

However, so far, no good cases have been found to testify for this assertion. But there had been found some other cases acknowledging the disadvantageous effect of this star. JACQUELINE KENNEDY has the fixed star positioned in a link-up with Sun = Mars/Saturn. It is known that she had several miscarriages, that one child only lived one day, and that her husband was assassinated, - and it is doubtful how her marriage with ONASSIS will fare.

MUSSOLINI had Praesepe in conjunction with Sun and Mercury and in semisquare to Uranus. This configuration points to his tie-up with the revolutionary movement and to his brutal ways, as well as his ignominous death by execution and hanging.

A female, born 27.7.1887, (see "Direktionen", Fig. 75) was badly injured by a wheel flying off from a truck. She had a fractured thigh, severe injuries to her intestines and head injuries, and she died in spite of immediate medical attention. At this time, Pluto s was passing over her Sun conjunction Praesepe complex.

The former Party Secretary of the Social Democrat Party, Dr. SCHUMACHER, came out of the First World War as a seriously disabled soldier. He always needed assistance when taking his place behind a lecturn. He had Jupiter in conjunction with Praesepe, but in square to Saturn.

The philosopher Dr. KARL JASPERS was smitten by illness all his life. Praesepe was in conjunction with his Ascendant.

32. NORTH ASELLI, 6°50' Leo,
Gamma and Delta Cancri,
33. SOUTH ASELLI, 8°01' Leo,

outflank the stellar cluster "Manger". This might be the origin, as seen in nativity pictures of the birth of Christ, of the positioning of a donkey always behind the manger. Both these fixed stars correspond to the effect of Sun and Mars. They are therefore to count as a positive influence, if in conjunction with Ascendant, Meridian or stellar bodies of the same character, but especially if grouped with Sun and Mars. These people will have an aggressive nature, and will not take insults easily. They may, by their own lack of caution, or by being dare-devils, put their life into danger; and they will not hesitate to use brutal and violent means. When the Aselli are with the Ascendant, danger by large animals (horses, bulls) may exist.

Uranus linked with the Aselli will indicate sudden tragic fate, such as accidents, falls, and emotional upsets. ELSBETH EBERTIN recorded in 1928 as follows: "I could see the effect in some horoscopes myself. Graefin STACHWITZ, born 29. 9. 1872, (Uranus in 5°08' Leo) was found murdered in Berlin. The sister of a Professor whom I knew, born the same day but at a different hour, was sent to a mental institution. Another lady I knew, born 29. 9. 1872, received severe injuries to her back when she fell from a ladder.

The former chief of S. S., HEINRICH HIMMLER, had Mars in 7° 57' Leo, and at the same time linked with MC and in Saturn/Pluto. He was the murderer of countless human beings and found his end by committing suicide.

The double murderer HELM (epa-101) had Mars positioned 5°51' Leo. He had 15 previous convictions. When in a car with a detective he grabbed a pistol from the brief case of one of the policemen and, in spite of his shackled wrists, he shot the policemen, found the key to his handcuffs and escaped. His sentence of death was commuted to life imprisonment.

The strong Mars influence, emanating from 5-8° Leo, must not necessarily be thought of as a destructive element. If a person is able to sublimate this Mars energy, then an extraordinary output of achievement is possible. Therefore it is not surprising to find, on this particular degree, the natal Sun of HENRY FORD, the natal Moon of LENIN and the natal Moon of MARLENE DIETRICH.

34. KOCHAB, 12°24' Leo, Beta Ursa minoris, 2nd magnitude,

meaning in Arabic "Buck", in "The Small Bear" or respectively on the back of the "Little Dipper". With an afflicted Sun, suicide is indicated. This star has not been researched much, and therefore it is only mentioned here shortly.

35. SERTAN (Accubens), 12°56' Leo, Alpha Cancri, 4th magnitude,

main star of Cancer, in "The Scissors of the Crab", has a Martian nature, and a strong blending in of Saturnian influence, conveying an unbalanced and "jumpy" nature. If the cosmogram is a disharmonious one to people affected by it, and especially if Sertan is in conjunction with Mars, Sun or Uranus, this will make for an unsettled mentality and helplessness. If in conjunction with Saturn, major disappointments in life, ordeals and trials, mental suffering, excitements, loss, opposition, disputes and deception are indicated. According to ELSBETH EBERTIN's experiences, this is the case especially if these persons are in positions of social standing or are politically active.

36. DUBHE, 14°27' Leo, Alpha Ursa majoris, 2nd magnitude,

in Arabic "Bear", main star of "The Great Bear" is credited with the destructive power of Mars, working itself out particularly in mundane maps, in a nasty way, if this fixed star is in conjunction with Saturn. Dubhe is positioned in conjunction with the Moon in the natal chart of MAO TSE TUNG, and with Saturn in the chart of HITLER.

37. MERAK, 18°41' Leo, Beta Ursae majoris,

in "the Side of the Great Bear" - being the second largest star of this constellation - has a Mars nature. Merak is of importance in a natal chart, if the sign Leo is well tenanted, and other configurations give a clue to love of command and domination. This fixed star is credited with increasing the power of the native to get on in life, and this especially in conjunction with Sun, Mars or Pluto.

Merak in conjunction with the Sun is found with air-ship pioneer Dr. ECKENER, H.P. BLAVATSKY, the founder of the Theosophical Society, and with scientist Dr. ALFRED ADLER, the founder of Psychology of the individual. He had a planetary picture of Sun = MC = Mars/Jupiter. A conjunction with Mars is to be seen in nativities of dictator ROBESPIERRE and spy MATA HARI.

38. RAS ELASED AUSTRALIS, 20°00' Leo, Epsilon Leonis,

the Southern star in "the Head of the Lion" is credited with properties such as conveying higher spiritual gifts of the Logos to those men who are able to conceive them. In primitive natives, this fixed star may become a danger, if bound up with a poorly placed Saturn or Neptune. Severe psychological depressions are indicated, and possibly even suicide. In conjunction with a "strongly" placed Mars, the Lion's Head will make for feverish diseases and. if conjunct with Uranus, dangers of accidents are present.

BISMARCK had this fixed star positioned exactly on his Ascendant; this fact as well as the part played by other configurations gave the chancellor his mental superiority.

The successful industrialist GOTTLIEB DUTTWEILER has the fixed star linked with Sun and MC. MAHATMA GANDHI's map shows a conjunction of Moon square Jupiter. Significant for her acting "with all her heart" is the conjunction of Ras Elased with Venus square Moon for film actress GRETA GARBO. The designer of the WANKEL engine had Mercury in conjunction with this fixed star. A peculiar picture is formed with this fixed star, in the chart of RAINER MARIA RILKE. Ras Elased in conjunction with Uranus opposition Saturn = Jupiter/Pluto. Chencellor Dr. ADENAUER had this fixed star in conjunction with Uranus at his birth, and it is significant in this latter case that Uranus with 92 distance value is "Lord of Tension" in his cosmogram.

39. ALPHARD, 26°36' Leo, Alpha Hydrae,

main star of "Water Serpent Hydra", in Arabic El-Ferd, "the one who stands alone", is of Saturnian nature. However, there is a measure of influence of Venus and Neptune. This combination is of disadvantage in most cases. Particularly matters connected with "poison" are accentuated badly, e.g. - blood poisoning, murder by poison, attempts of poisoning, poisoned hatred in women, gas poisoning, danger to life by wrong use of drugs and over-indulgence of good living, smoke inhalation and danger of suffocation, snake bite, bites by poisonous insects, or bite by dogs with rabies. Danger is marked if either Sun, Moon, Mars, Neptune, Ascendant or Meridian are linked up with Alphard. A "poisoning" is also possible in a relationship between man and woman in marriage. This is especially true in a male cosmogram.

The wife-murderer of Breslau FRANZ GEPPERT had Uranus in conjunction with this fixed star. He was sentenced to death and executed. The same configuration is found with LEO ERICHSEN. He raped girls while they were hypnotised, and he was put into jail. A real monster was the mass-murderer DENKE. He lured unsuspecting people into his house and, after murdering them, he "ate" them! ELSBETH EBERTIN, who investigated several cosmograms of his victims, found Alphard in conjunction with Sun or with Malefics. A further example is given in Kaiser FRANZ JOSEPH I of Austria. Born 4 hours before a Solar eclipse, he had Sun, Moon and Saturn in degrees 22°-25° Leo in his chart. He was capable of keeping together the monarchy on the Danube, in spite of a severe shaking of wars and dramatic events in his family.

On the material plane, Alphard is nearly always unhelpful, though if the native is able to conceive it in the spiritual sphere, the Saturn-Neptune combination may give enlightenment. ELSBETH EBERTIN had Alphard conjunct MC. Maybe this indicates that she was busy investigating the tragic lives of many people. Certainly, in this way, she gained a great insight into her particular work. Her death by suffocation however is significant for the link in her map between this fixed star and her MC. The physician and scientist of metaphysics Dr. GEORG LOMER has this fixed star in conjunction with Uranus.

40. REGULUS, 29°08' Leo, Alpha Leonis, 1st magnitude,

the "Little King in the Heart of the Lion", called the Royal Star it may convey royal properties, noble mind, frankness, courage. The importance of this fixed star is accentuated by the nearness to the ecliptic. Its effect is in the best sense that of Jupiter and Mars.

On the Ascendant, it will give a courageous and frank character, especially if in conjunction with Sun, Moon, Jupiter or Mercury. Positioned on the Meridian, Regulus will raise the native to high positions in life, positions far exceeding the environment the native was born into. According to tradition, this configuration will bring with it connections with rulers, honourable people or famous people, if the cosmogram as a whole points to this possibility. Jupiter conjunct Regulus is one of the very best configurations for success. People like this can reach high positions, positions of trust, preferment and favours; fortunes. Riches and power could fall to these natives easily.

ELSBETH EBERTIN did have this fixed star near the Meridian, but the Jupiter link was absent. Therefore, in spite of diligence and hard work, she could not grow rich. She always had to live plainly and modestly. An old saying goes that Regulus in the 10th house "makes astrologers for Kings, people in high positions and noble men". Naturally not everyone who has this configuration will make a "Royal Astrologer", even if he has the necessary knowledge and capabilities. But with Frau EBERTIN, this rule was borne out. As a graphologist, she already had the job of reading the handwriting of Kings and Counts, and she was given credit for it. During the First World War, her book "Royal Nativities" was published. This was followed up with further far reaching contacts, especially with the former King FERDINAND of Bulgaria and his family. She was commissioned to set up the horoscope of the later Crown Princess of Sweden. She was invited by Royalty in Southern Germany and consulted by them. On the other hand, she had many contacts with ruling industrialists.

Regulus culminating is not only a good omen for a military career, but also for careers connected with the public, such as lawyers, civil servants, bankers and the clergy, especially if Regulus is also in conjunction with one of the lights (Sun or Moon) or with a benefic stellar body.

ELSBETH EBERTIN studied historical events with the planetary configurations in each case and came to the following conclusion: "When Mars or Saturn are in conjunction with Regulus especially exciting events are always recorded e. g. assassination, coup d'état, revolution, revolt, demonstration and similar events." The investigating Frau EBERTIN had a long list of comparable events, and only two of these will be picked out here:

On 20th March 1890, with Saturn conjunct Regulus, BISMARCK was forced to step down from the position of Chancellor.

In November 1918, the same configuration, plus a link with Mars and Uranus, led to the overthrow of "crowned heads" and military leaders as well as to the Revolution in Germany. This event was also recorded by Frau EBERTIN in her "Sternblaettern", and in "Koenigliche Nativitaeten" she made hints to this effect. The former President of the French Republic, POINCARE, had a natal Sun-Regulus conjunction, and he was overthrown after the First World War. The German "prince of poets" WOLFGANG GOETHE had in his nativity the "Royal Star" in conjunction with Mercury and in square to Pluto, indicating his far reaching influence (as GOETHE said "the poet should keep friendship with Kings" (= German: "Es soll der Dichter mit dem Koenig gehen").

It is most interesting to note that the French King LOUIS XIV had Regulus conjunct MC and associated with Pluto = Mars/Jupiter. This configuration corresponds exactly to "the Ruler of the rigid Absolutism", who said about himself "Just as complete perfection and all virtues are in God, so is all power of the single people bonded in the person of the King" (see German book on Pluto, page 170). The former President of the German Bundestag, Dr. HERMANN EHLERS, representing the second highest position of the Government, had Regulus conjunct with Mars and also tied up with the MC.

41. PHACHD, 29°41' Leo, Gamma Ursae majoris,

the third largest star of the "Great Bear", also called Phekda ("thigh of the Bear") corresponds to Mars, with added touch of Uranus and Neptune. As it is so very far distant from the ecliptic, its influence perhaps is not felt very much. In conjunction with planetary malefics, it is said that this star is a possible cause of "a great blood bath". In conjunction with Neptune, and provided relevant configurations with Moon, Venus and Mars are also present, it is an indication of a pathological sex nature.

The power of this fixed star is a distinct one in mundane astrology. ELSBETH EBERTIN noted that when the assassination in Serajewo of the Austrian heir to the Crown took place, Mars had just transited this fixed star. When the Justizpalast in Vienna was overrun, 20-24 July 1927, Mars transited Phachd, and in other places too, riots took place.

42. ALIOTH, 8°09' Virgo, Epsilon Ursae majoris,

in "the Tail of the Great Bear" is not supposed to have a strong influence on account of its great distance from the ecliptic. However if Alioth is felt to make an impact, it is Mars like and of a destructive kind.

Associated with the Sun, suicidal thoughts will be eminent in female cosmograms. Joined with the Moon, it will be an indication of a possible danger during pregnancy and delivery. Amalgamated with nonpropitious planets, it will indicate fiascos, bad fortunes within the family circle or within the profession or disappointments brought about by friends. It is advised not to include these findings in a prognosis, as these records were not borne out by later research.

The Austrian philosopher Dr. RUDOLF KASSNER, may serve as an example. He has in his nativity Mercury 6°59' Virgo, near Alioth and in square to Mars. Early in his life, he was struck by infantile paralysis. He had to use crutches all his life. Yet, in spite of this handicap, he travelled around the world. He has a reputation as "one of the last interpreters of the conscience of Western culture. A really negative outcome is shown in the cosmogram of mass-murderer HAARMANN. In his birth chart Alioth is situated in conjunction with Uranus = Sun/Saturn = Saturn/MC.

In the Pluto Book (18), figure 23, there is the birth chart of a child, suffering from an incurable congenital impairment of his metabolism. In this chart, Pluto is conjoint with the fixed star and in square to Moon = Venus/Neptune = Saturn/MC.

43. ZOSMA, 10°35' Virgo, Delta Leonis,

on "the Back of the Lion", has a Saturn-Venus nature. Its reputation is that of giving an alert mind, but also inclination to melancholic moods. Conjunct with the so called "malefics", danger by poison and disease of the intestinal tract are indicated. These interpretations have to be considered with utmost caution and restraint.

In our archives, we found these examples: Female, born 6. 6. 1881: Zosma in conjunction with Uranus = Moon/Neptune (women's diseases) was her radical pattern. Her uterus was almost totally undeveloped. She frequently missed her periods, and she died of carcinoma of the oesophagus. Male, born 29. 1. 1901: Zosma was in conjunction with Mars = Sun/Saturn. He suffered from carcinoma of the prostata.

44. MIZAR, 14°52' Virgo, Zeta Ursae majoris, 2nd magnitude,

on "the Tail of the Great Bear", and as a companion and only 11' distant is situated the star Alkor (Arabic for "black horse"), also called "the Little Horseman". The latter is of 4th magnitude and only people with excellent eyesight can distinguish it as a separate star. Supposedly, Mizar portends a Mars nature. The reputation of Mizar, if it is in maximal position in a mundane map, is that of being connected with fires of catastrophic extent and mass calamities. In personal charts, Mizar is not helpful if conjoint with "bad" planets. It is not wrong to assume that, besides these handicaps, artistic emanations can also be attributed to Mizar. It could be of relevance to MOZART, who had at his birth Mizar in conjunction with the Ascendant (this is only if the supposed minute of his birth is recorded correctly). MARLENE DIETRICH also had Mizar conjunct Ascendant. One should give a positive reading to Graf ZEPPELIN regarding his creative spirit, but also a negative one regarding the catastrophes in which several air-ships perished.

The Blood Wedding of 23/24 August 1572 when, throughout France, nearly 20,000 Hugenots were massacred, is very much in character with Sun conjunct Mizar. In Paris alone, 2,000 people were murdered.

OSWALD SPENGLER, the writer of the book "Untergang des Abendlandes" has Mizar exactly on the mid-heaven.

45. DENEBOLA, 20°57' Virgo, Beta Leonis,

in "the Tail of the Lion" has a Uranus nature and it is supposed that, in mundane horoscopes, major catastrophes are triggered off by it. Depending on position and aspect to other stellar bodies in a personal cosmogram, either preferment or fall are credited to this fixed star. Found on the Ascendant and especially if in company with Mercury, a quarrelsome nature, with a liking for legal action, is attributed to Denebola. It also could mean that this star is the cause of very exciting events. Badly placed in a map, with Mercury or Uranus, mental diseases - and these mostly incurable ones - are indicated. Fine aspects, however, will further work connected with matters of reform and progress.

Denebola was in a conjunct position with the Ascendant, when one of the first giants of the ocean, the TITANIC, was launched, and nearly all her passengers and her crew drowned with her. About 1,500 lives were lost. HITLER, when assuming power had the same configuration, and millions of people suffered under him.

The German explorer WILHELM FILCHNER first became known with his "Ritt durch Pamir" (= "Through Pamir on Horseback"), which he did with two horses and without any companions. He was taken captive in 1934 in the piratical State Tungan in East Turkestan. This coincided when Sun directional came in opposition to Neptune, and radical Sun became conjunct with Denebola. He was released only in 1937.

Mass-murderer PETER KUERTEN had Denebola conjunct Uranus = Moon/Neptune and square Mercury, identifying him as an impulsive, unpredictable and dangerous man. In the natal chart of Italian dictator MUSSOLINI we find the same tie-up with Uranus, being in semisquare to Sun and Mercury. A likelihood of elevation and fall are clearly prominent.

CONRAD HILTON, the "largest Hotel Industrialist of the World" had an unusual rise. At his birth, Denebola was conjunct his MC square Mercury = Mars/Pluto. The German industrial magnate FLICK has the fixed star united with Uranus = Sun/Neptune = Moon/ Saturn = Saturn/MC. He was taken prisoner after the last World War. This happened when directional Sun passed over radical Uranus. FLICK lost 75 % of his industrial capacity by forced dismantling and expropriation; he was however released in 1950 prematurely, and by skilful transactions he has been able to recreate an even larger combine. It is known that his sons caused him a lot of trouble.

46. BENETNASH, 26°08' Virgo, Eta Ursae majoris,

the last star in the "Great Bear", meaning "hired mourners". If the influence of Benetnash is exercised, an influence of a Mars-Uranus-Saturn nature is present. Experience has shown that many human lives are to be mourned. REINHOLD EBERTIN made a survey of this fixed star covering centuries when associated with transits of the major planets over this degree. The results have been recorded in the 40th. "Jahrbuch fuer Kosmobiologische Forschung 1969" (= Year Book for Cosmobiological Research 1969). In accordance with adopted belief of ancient times, this fixed star is supposed to be bound up with the realm of the dead and is therefore associated with death and mourning. In an important position in a mundane map, Benetnash will claim human lives in calamities such as mine accidents, collapse of houses and bridges, mountain slides, earth tremors and catastrophes caused by weather.

Uranus was in exact conjunction with Benetnash at the end of July 1968. This time was marked with extreme tension between Prague and Moscow, followed later in August by the Russian take-over of the CSSR. Tank crews occupied Damascus broadcasting station. At the same time there was a rebellion in Yemen, racial strife flared up in Indiana, USA, the Czech reformers in Schwarzau, on the Theiss fought for the freedom of their nation on the 30th July, with the Russian political big wigs, the Soviets widened their "manoevres" in Poland. In the DDR, at the same time, there began the mass movement of the Viet Cong. Bloody riots took place in Mexico. And at least 100 human lives were lost when a volcano erupted in Costa Rica.

FRIEDRICH NIETZSCHE

born 15.10.1844 in Roecken, 10.07 a.m.
died 25. 8.1900 in Weimar.

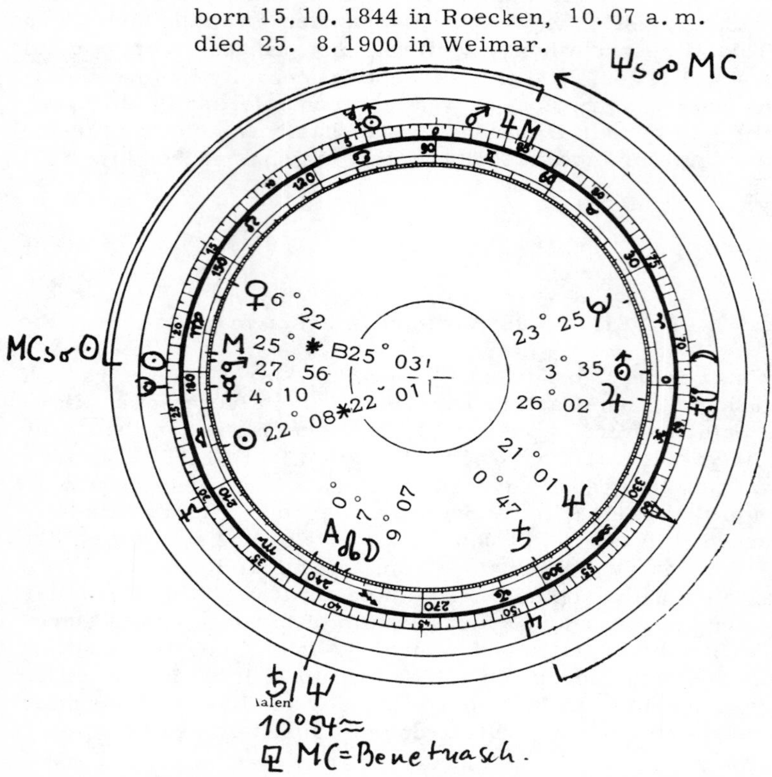

Sun is in conjunction with Spica and MC and Benetnash in this cosmogram. Amongst other properties, Spica has a reputation for helping scientists, writers and artists to succeed. When MC s reached the conjunction of Sun with MC 1869 he became Professor. One year later, he was made Ordinarius (a German University title). During this time "Empedokles" was created. When Neptune s reached the opposition to MC with Benetnash he experienced his climax - a climax of his syphilis of purely neurological origin. On the death of NIETZSCHE Saturn s had come to MC with Benetnash. (Please note that directions by measure of Solar Arc (s) are marked in the 90° sphere).

About the 3rd November 1968, Mars transited Benetnash. These days were marked with the loss of over 100 human lives in floods and high water in Italy, and in Jugoslavia there was an earthquake.

On 19th September 1968 Sun transited Benetnash. Many houses collapsed as a result of an earthquake in Caracus on the 20th. In addition, many accidents were reported.

Tied together with Saturn in a natal cosmogram, Benetnash is supposed to lower vitality. If both Mars and Uranus are poorly placed, the death of the native is said to be a sudden one.

The philosopher FRIEDRICH NIETZSCHE had, in his birth chart, a Benetnash MC conjunction, in semisquare to the disease axis Saturn/Neptune. He was pronounced to be in a far advanced stage of senility when directional Meridian came in square to Saturn/Neptune.

The former Lord Mayor of Leipzig, Dr. KARL GOERDELER, became Minister for Price Control under HINDENBURG. Under HITLER he was a member of the underground and was designated to become German Chancellor after HITLER's death. But in 1945 he was executed. On the day of his birth, Benetnash was in conjunction with Uranus = Mars/Saturn. When, by direction, this group reached the square to Mercury, he was killed.

Significantly in the case of mass-murderer and rapist HAARMANN, we find Benetnash in conjunction with Venus = Saturn/Uranus.

47. VINDEMIATRIX, 9°16' Libra, Epsilon Virginis,

"The Woman Winegrower" in the constellation of Virgo, rises not long before grape-harvesting time. On the Meridian or on the Ascendant this fixed star with its Saturn-Mercury nature is said to help mental concentration and to promote the type of native who engages for architects and businessmen. Tied up with Mars, it results in increased power of thought, tending sometimes to sarcasm and polemics. A person of this description was the former MAXIMILIAN HARDEN who became known by his needling criticism during the last years of the German Kaiser; e. g. he published political material at the time of the EULENBERG - affair. His natal

Mars is conjunct Vindemiatrix. The philosopher LUDWIG KLAGES also has his Mars in the same degree of the Zodiac, and he was known by his severe comments. It is due to KLAGES that graphology was accepted in German universities. FRIEDRICH EBERT, first President of the German Republic, had the same configuration.

Both HINDENBURG's and GANDHI's natal Sun were in conjunction with this fixed star; the difference between the General and the peace loving revolutionary has to be recognized in the position of Sun and the complete cosmogram.

A very significant Vindemiatrix conjunct Mercury square Pluto is to be seen in the chart of actor and show master VICO TORRIANI, who is able to carry away millions of viewers who watch his show "Der goldene Schuß", a popular entertainment on television in Germany. His talents were noted for the first time in 1946 when directional Mars formed a square to Mercury and an opposition to Pluto.

Badly placed, and this especially so if in conjunction with Saturn or Neptune, this fixed star is said to cause a tendency to depressive moods, scepticism and a distrustful suspicious nature. Tied up with Mercury, Vindemiatrix - if also otherwise badly placed - will lead to nervous irritability. In conjunction with Mars, a danger of injuries is indicated: MAXIMILIAN HARDEN suffered severe injuries when an attempt on his life was made, FRIEDRICH EBERT died after an unsuccessful surgical operation.

48. ALGORAB, 13°12' Libra, Delta Corvi,

in "The Crow", has a Saturn-Mars nature and is said to be of hindrance especially when other Saturnian influence is predominant. Delays and restraint are indicated to come about by fiascos, losses, wrong handling of matters and enmity in general.

The cosmogram of Kaiser FRANZ JOSEPH I is a good example. He has Algorab on his Ascendant, and his life was full of misfortunes. A conjunction with Sun or Moon is not good either, if the lights are afflicted.

Algorab in conjunction with Venus = Saturn/Pluto is to be seen in the birth chart of Princess MARGRET ROSE of Britain. Her first love affair turned sour, and her marriage is generally known as not being entirely successful.

The poet GEORG TRAKL, "the lyricist of the Dying Occident", had in his birth chart a configuration of Algorab = Uranus = Neptune/ Pluto = Sun/Moon. He suffered increasing mental anguish leading to his becoming a patient in a psychiatric clinic. He took his own life there by taking poison.

A female, born 13.3.1903, (see "Direktionen", page 100): Algorab is in conjunction with Mars = Sun/Saturn = Moon/Saturn. Her husband suffered from multiple sclerosis and died soon after marriage. She also was stricken with this disease, picked up by contact with him, and her illness was found to be incurable.

Male, born 28.1.1893, had Algorab conjunct with his Saturn. Due to political denunciation, he had fallen into disgrace following the war, and lost his position. Only years later was he rehabilitated. Lowering of his status caused him long lasting gastric troubles.

According to ASBOGA (13), if Algorab is in conjunction with either Sun, Moon or one of the "Malefics", it has a reputation for accidents and injuries which are difficult to avoid.

49. SPICA, 23°08' Libra, Alpha Virginis,

situated in "The Wheat Sheaf" of constellation Virgo. It has, on account of its size and nearness to the ecliptic, some importance. Spica has a Venus-Mars nature. To Spica are ascribed honours and fame. Spica is of marked good influence for scientists, writers, artist-painters, sculptors and musicians. If Spica is on the Ascendant or with the Meridian and in conjunction with Jupiter and Venus, promise is given even to people of humble origins to enjoy protection, preferment and even wealth.

In conjunction with Mercury or Venus and positioned on the Meridian or Ascendant, Spica will give artistic skills, ability to draw, musical talents and a good sense for literature and the sciences.

However, if Spica is placed in "angular houses" and conjoined with Saturn, Mars, Uranus, Neptune or Pluto, and if these planets are afflicted, a rise followed by a downfall with tragic ending could be the result. In 1923 when, as a result of inflation in Germany many people became poverty stricken, Spica was bound up with Saturn.

In the personal cosmogram, Spica will mean refinement. Such people will have a noble bearing. Increased erotic tension, usually given with a Venus-Mars interpretation, is not the case with Spica, but a tendency to sublimate these powers and turn them into artistic and creative channels is indicated.

VEHLOW (12) pointed out that the First World War pilot IMMELMANN, well known at that time, had grouped together in his map Spica, Sun and Uranus, all on his Ascendant. It would be interesting to investigate whether other pilots had positions as exalted as this in the "airy" sign Libra.

Mention should be made here of the significance of Mars near Spica in the case of well known aviators Captain KOEHL in 20° and ERNST HEINKEL in 22°15' Libra. KOEHL took off on 12. 4. 1928 to Labrador via Ireland. This was the first ever trans-atlantic flight. In those days, this was an extraordinary achievement. Needless to say, on this day Sun was in 22° Aries, in opposition therefore to Spica and radical Mars. ERNST HEINKEL was a designing engineer and pilot. By 1910 he had designed the first double decker. Later he developed start by catapult. He made his name with eight international records which he gained with his plane HE 70 "Lightening", the first fast plane known in the world.

Scientist/Inventor EDISON had Lunar Node conjunct Spica and in opposition to Pluto and semisquare Jupiter.

Amongst the literary celebrities is found HERMANN HESSE. He has Spica conjunct MC square Venus. THOMAS MANN has Spica conjunct Jupiter square Moon. VICKI BAUM, well known novelist, has Spica conjunct Mars. The composer BRAHMS had Jupiter united with Spica.

50. ARCTURUS, 23°32' Libra, Alpha Bootis,

"Bear leader-guardian", meaning observer of the "Great Bear", or "The Saucepan" or "Big Dipper" respectively. Main star of constellation Bootis (= driver of oxen), has a Jupiter-Mars nature, and a reputation of achieving "justice through power". It therefore makes the native belligerent and quarrelsome, especially if attached to Mars and Jupiter by conjunction. A really go-ahead and enterprising spirit is here the rule, as indicated by Jupiter-Mars. Lasting success is promised if further good aspects are present. If critically aspected, the good influence will be hampered or made into a real handicap. If involved in legal action, such a native may lose all.

Between Spica and Arcturus there is a minor difference in longitude but a very large difference in latitude. A blend of mutual influences is frequently given. Although Spica is very much larger and will dominate, it may get more positive character by its blending with Arcturus.

51. THE SOUTHERN CROSS, 11°11' Scorpio, Alpha and Beta Crucis,
(Unfortunately not visible for observer in the Northern Hemisphere)

is a magnificent arrangement of stars. The Jupiter nature of this fixed star is especially impressive if its latitude is the same as that of the Ascendant. Then this star is supposed to be credited with intuition, a grasp for the inner nature of one's fellow man, a preference for occult studies, the gift of successful investigation of the hidden side of things, an inventive mind, a deeply religious nature connected with mystical and theosophical interests. ELSBETH EBERTIN had an exact conjunction Ascendant-Southern Cross, and the reading given fits perfectly in her case.

This interpretation of the Southern Cross is especially true if the sign Scorpio is in favourable cosmic circumstances in the chart and the negative portents of Scorpio are eliminated thereby.

The noted psychiatrist KRETSCHMER, known for his book "Koerperbau und Charakter", has a Mercury position of 10°32' Scorpio.

ERICH KAESTNER, "the most influential and well known German author", a shrewdly observing moralist and poet, has in his birth chart The Southern Cross conjunct Jupiter in 10°12' Scorpio and linked with the Ascendant = Pluto/Lunar Node = Sun/Venus = Sun/Mars.

52. GEMMA, 11°32' Scorpio, Alpha Coronis borealis,

the "Jewel" of the "Northern Crown", has a Mercury-Venus influence with some properties of Moon and Neptune. Contrary to the Southern Cross, Gemma has a high Northern declination. Both stars differ in position by variance in Rectascension, a difference in position stretching over more than a quarter of the celestial global arc. Measured in ecliptical longitude, it is difficult to separate the mutual influence of each of these stars.

Because of its Mercury-Venus properties, Gemma, if in good position and found on the Ascendant, is credited with a liking for literature, art and science, artistic talents, and also success in trade and commerce.

If conjunct with Sun and Moon and well placed in a map, Gemma promises honours and preferment. According to tradition, Gemma is said to be associated with "The Lord of the House of Death", and if connected with Neptune and Mars, infectious diseases and poisoning are indicated.

HANS HEIZ EWERS had a conjunction of Sun, Mercury and Gemma; he is known to be a poet who portrays the grotesque and the hideous, monstrous and terrible, blended with eroticism and mysticism (Scorpio). His half sum Venus/Mars is only 2° away from this grouping. The titles of his novels "Vampire", "Hunter of the Devil", "The Horror", "Alraune" are significant indications of his work.

The cosmogram of the former leader of the German party (SPD), Dr. KURT SCHUMACHER, shows Gemma in conjunction with Mercury in 11°05' Scorpio.

53. THE SOUTHERN SCALES, 14°23' Scorpio,
Alpha Librae,

is a double star. Its brighter partner is of Mars character with a weak blend of Saturn. The unhelpful influence of this fixed star is said to be noticeable especially in natives who were born during the night. A bad omen for the health of the native is given if Sun or Moon is in conjunction with The Southern Scales. It has been noted that it is particularly disadvantageous to have Saturn and Neptune tied up with this star. Even the conjunction with Jupiter is credited as being adverse. VEHLOW (12) has recorded "confiscation of possessions during times of war". A credit given to both The Southern Scales and The Northern Scales is that they bestow an immortal name. However, this comes about more by tragic circumstances than by well noted success. Conjunct the Ascendant, the fixed star could mean a rise in fortunes, followed by a down fall. There is also an indication of danger coming about in connection with water.

The following examples will show how tragic fate overcame these particular natives: MARIE ANTOINETTE, the unhappy Queen of France who was executed.

BERND ROSEMEYER, popular racing ace, died in an accident when only 28 years old. His cosmogram shows Ascendant in 13° Scorpio and tied up with Mars/Pluto. This particular configuration was triggered off by ROSEMEYER's MC, advanced by measure of Solar Arc.

The former Minister for Propaganda and "mass deceiver" Dr. JOSEF GOEBBELS had his radical Mars 13°33' Scorpio in conjunction with this fixed star. His end came about by suicide in the bunker of the German Reichskanzlei.

The physicist BURKHARD HEIM lost his eyesight when engaged in experiments. He had Saturn 14°11' Scorpio = Uranus/Pluto.

The mother of a Thalidomide baby (in Germany this drug was marketed under the name "Contergan") had Jupiter conjunct with the fixed star, but also in half sums Mars/Neptune = Mars/Saturn = Moon/Pluto (Pluto Book, page 111). The example "Inge Lehmann", as shown in the book "Lebensdiagramme", page 104, also had Jupiter conjunct with this fixed star and = Sun/Neptune. She gave birth to a child which was not able to develop in a normal way and this baby had to live in an institution.

54. NORTHERN SCALES, 18°40' Scorpio, Beta Librae,

has a Jupiter-Mercury nature and therefore positive properties. Tied up with Meridian and Ascendant or with well placed stellar bodies, the Northern Scales are credited with helping the native to gain honours and distinction. When connected with Mercury they are supposed to make the native studious. In good position, the fixed star will arouse, above all other things, spiritual and mental forces. The conjunction with Sun, Moon or Jupiter favours civil servants, lawyers and scientists.

WOLFGANG VON GOETHE, "the mystic", had the Northern Scales conjunct Ascendant. His map also showed a square to Uranus.

The birth chart of MUSSOLINI, the Italian dictator, shows the fixed star in conjunction with Ascendant = Sun/Pluto (succeed by power of own personality) = Sun/Jupiter (with success).

MAHATMA GANDHI, who freed India from British rule by non-violent action, was born when Mars was 18°25' Scorpio and in opposition to Pluto. GANDHI became known world-wide and was admired by millions, but in spite of all this, he was assassinated.

The South African Premier VERWOERD had the fixed star united with MC and semisquare Jupiter. He reached the top position in the State, but the first attempt on his life brought him injuries and the second attempt brought him death.

Prof. Dr. HANS DRIESCH, a widely travelled visiting lecturer, whose books are known all over the world, had the fixed star in conjunction with Moon. It is interesting to note that he was the first University Professor in Germany who had the courage to write a prefix to an Astrological major book (v. KLOECKLER: "Astrologie als Erfahrungswissenschaft"- Astrology as science of experience).

The great English Statesman WINSTON CHURCHILL had the fixed star in his natal chart conjunct with Mercury = Uranus/Pluto.

55. UNUK, 21°23' Scorpio, Alpha Serpentis,

in "the Neck of the Serpent" shows its properties as a Martian force combined with Saturn, and often is really dangerous and destructive. There will be chronic diseases, which are not easily detected, - these will result in a weakened state of health, and operations will be necessary. There will be accidents. This is especially true with Uranus or with Neptune. The conjunction of the latter with Unuk will make the native prone to infectious diseases and also to poisoning. If with the Meridian and in conjunction also with "Malefics", Unuk is damaging to the social position. Often there will be many difficulties in the professional life.

SUKARNO, the deposed dictator of Indonesia, had at his birth Unuk conjunct Lunar Node = Mercury-Jupiter/Neptune.

The "proposed Reichspraesident" Dr. GOERDELER, executed by HITLER, had his radical Moon conjunct Unuk in 20°40' Scorpio and in opposition to Neptune = Sun/Pluto.

VERA BRUEHNE, accused of having helped in the killing of her lover, Dr. PRAUN, was sentenced to hard labour; her birth chart shows Unuk conjunct Ascendant = Neptune/Pluto = Saturn/Pluto = Sun/Venus.

Girl, born 14.11.1939, took her own life by jumping from the Cologne Cathedral on 14.11.1959. She died immediately. Her cosmogram shows Unuk in conjunction with Sun and in opposition to Uranus.

Woman, born 17.10.1896, was suffering from kidney trouble for many years. Her birth configuration shows Unuk in conjunction with Venus = Sun/Neptune = Saturn/Uranus.

Another woman, born 6.11.1896, suffered long drawn-out back troubles, and her chart shows Unuk conjunction Saturn = Sun/Moon = Moon/Uranus = Mercury/Pluto.

An ammunition ship of the US Navy exploded on 19. 5. 1950, when Unuk was conjunct Ascendant (29 dead, 300 severely injured).

An aeroplane accident near Zurich took place on 4. 9. 1963 when Unuk was conjunct Mars square Saturn.

56. BETA CENTAURI, 23°06' Scorpio,

is endowed with a strong Venus characteristic with a touch of Jupiter blended in. Not visible to observers living in the Northern Hemisphere, it is a star of Southern latitudes. If this fixed star, in its degrees of ecliptic, is in connection with Mercury or on the Ascendant and in conjunction with a "Benefic", it can be taken as a promise for happiness and success. Tied up with Venus however, this star will make for pronounced sensuality; if, in addition to this, it is tied up with Mars, Saturn or Neptune, then the native may be the subject of gossip and scandal.

A significant example is BARBARA HUTTON, the Woolworth heiress, who has married seven times so far, and on an average rearranged her marital affairs every 4 years. This has given the gossip columns of a number of papers ample material. Her natal Sun in 22°04' Scorpio is in conjunction with Unuk.

Besides this, however, no other corresponding cases have been noted.

57. BUNGULA, 28°51' Scorpio, Alpha Centauri,

is also called "Toliman", as the former, like the previous star, this is only visible in the Southern latitudes. Bungula is exceptional amongst all the fixed stars, known up till the present, as it is the nearest to Earth. Its light "only" takes 4 years to reach us.

Also Bungula as well has a Jupiter-Venus nature, but there is the added peculiarity that relationship to female persons often seems spoiled, or an existing happy relationship is stricken by exceptional circumstances. Well placed, Bungula may help the native to gain a position of honour and power.

The main representative of English imperialism, Lord BEACONSFIELD (Disraeli), had a conjunction of Ascendant, Jupiter and Neptune with Bungula, and Venus was positioned with Beta Centauri.

The conjunction of Neptune with Bungula appears to be most disadvantageous in mundane maps. If one looks into past events, they always seem to tally up with disturbances, riots, periods of storm and stress and revolt against curtailment of freedom.

The world-acclaimed conductor HERBERT VON KARAJAN has in his cosmogram a conjunction of Bungula with Ascendant and opposition to Venus and Mars. WERNER HEISENBERG has a conjunction with Mercury=Mars/Jupiter, and the scientist HANS DRIESCH has a conjunction with Mercury square Jupiter.

58. AKRAB, 1°51' Sagittarius, Delta Scorpii and THE FOREHEAD OF SCORPIO, Beta Scorpii,

both in constellation Scorpio and positioned next to each other, are supposed to correspond to Mars character, with a blend of Saturn characteristics. Dr. KOCH is of the opinion that both these stars are "doubtful". The effect of these two stars has yet to be researched more fully. According to tradition, these fixed stars are credited with giving an ability to do research, and especially research into things of a particularly secret and hidden nature. For this to be, however, further influence of the sign Scorpio and planets Mercury and Uranus have also to be present; preferably these stellar bodies would be positioned on the Meridian or Ascendant and be near another stellar body pointing to mental capabilities. In lower types, however, a tendency to falsehood and treason can be noted. Both stars are a poor augury for material wealth. There will be difficulties, impediments, or loss, dependent upon the position these stars have in the cosmogram. In mundane maps, there is an indication of mass catastrophes, should either Mars, Saturn or Uranus be in the first degrees of Sagittarius, especially if in an angular position.

Confirmation of this has been recorded by ELSBETH EBERTIN: when in July 1927 Saturn transited this particular degree, the ghastly water catastrophe happened in Berggriesshuebel, when 150 human lives were lost; bloody riots took place in Paris; in Portugal there was a revolution; there was a revolt by Red Indians; and other political demonstrations were carried out.

Many years ago, I drew attention to January 1956, when again Mars and Saturn were to be conjunct in the same degree with this fixed star. The Mars transit was exact on January 16th. In my records were noted for this day: confrontation in South Italy, demonstrations in Bombay and a strike of 100, 000 laborers; 109 dead in battles in Algeria; a hurricane over the North Sea and the Baltic Sea; an earthquake in Ecuador: and in addition many minor catastrophes and accidents.

ELIZABETH TAYLOR, the film actress, has her Ascendant in 2°20' Sagittarius in square to Mars in 1°23' Pisces and in semi-square to Uranus. When it was found that she was not tall enough, her ambitious mother tried stretching exercises to make her gain height. When she was forced to learn horse-riding, she became exhausted by over-exertion, had a fall and injured her spine. All her life she has had a weak body and she is of delicate health. Important here, too, is the position of Pluto = Mars/Neptune.

SYLVESTER MATUSKA, the criminal, known for his activities regarding the great railway accident, had radical Mars in 2°46' Sagittarius, and another similar railway criminal (30. 6. 1950) had in this degree his Ascendant in square to Saturn.

A women fell from steps in her garden. This caused her to develop serious health trouble and from this cancer of the uterus developed. Mercury's position was in 1°30' Sagittarius.

59. ANTARES, 9°04' Sagittarius, Alpha Scorpii,

in the constellation of Scorpio, is - as its name Ant-ares says - of Martian nature but, to this Mars nature, powers of Mercury, Jupiter and also of Saturn are added. Antares makes people tough, belligerent and pugnacious. This is an important star for military personnel, and is said to convey mental alertness, strategic ability, and courage and to make dare-devils, especially if tied up with the Meridian, Ascendant, Sun or Jupiter.

If associated with Mars, courage is said to become foolhardiness, leading to increased dangers. Natives with this particular configuration have to be prepared at all times for sudden incidents and unforeseen events and potential accidents. According to tradition, Antares is of violent character and is credited with being significant for a violent death, either in battle or by process of law. On the other hand, danger may come about by fire, weapons or machinery.

Antares is also said to be unfortunate for the eyes, if in conjunction with Ascendant, Moon or Sun, and this has been proven to be the case by later researchers.

Antares was conjunct Ascendant in birth charts of GANDHI, who died bv assassination, General SCHLEICHER, murdered by HITLER, HELEN KELLER, born blind and known world-wide as a writer, and also of a woman, born 15. 1. 1921, who died by a car accident.

A conjunction with the MC has been recorded with boxer TEN HOFF and BORMANN, Minister under HITLER.

Conjunction with the Sun can be found in charts of the British Prime Minister CHURCHILL, Field-Marshall KESSELRING and the former Reichsmarschall GOERING.

Antares was in conjunction with the Moon of COPERNICUS, and bound up with the Lunar Node and Moon of the philosopher NIETZSCHE who, when only 45 years old, became an imbecile.

The Austrian "Kulturphilosoph" Dr. RUDOLF KASSNER was born when the fixed star was in conjunction with Mars. Because he contracted Infantile Paralysis, he had to use crutches all his life.

Male, born 15. 1. 1923, had prostate trouble when only 30 years old. This happened when Venus, radically tied up with Antares, was progressed by measure of Solar Arc, and thus reached the position of Pluto.

60. RAS ALGETHI, 15°27' Sagittarius, Alpha Herculis,

"the Head of kneeling Hercules", possesses a character of Mars-Venus with a slight blend of Mercury. If in good aspect, it is supposed to give much enjoyment and favours from women. However, if connected with unhelpful planets, it is said to cause much irritation, especially in connection with the female sex. But, by its very Mars nature, Ras Algethi corresponds to boldness and a drive to gain power. (Please note, in some text books and tables, the position of this fixed star is recorded with a difference of up to one degree.)

The fixed star was in conjunction with the Ascendant at the birth of the explorer and writer SVEN HEDIN and also of the story teller ANDERSEN.

ALBERT EINSTEIN, the atomic physicist, had a Moon conjunction with Ras Algethi. We also find, in the cosmogram of philosopher and physician ALBERT SCHWEITZER, a conjunction with Venus.

In the birth charts of WILLIAM BOOTH, founder of The Salvation Army, of physicist WERNHER VON BRAUN and of actress SABINE SESSELMANN, a conjunction with Jupiter is found.

61. RAS ALHAGUE, 21°42' Sagittarius, Alpha Ophiuchi,

in "the Head of the Snake Charmer", has a Saturn like character, and some of the undesirable Venus qualities are also present. Connected with these are Neptunian tendencies, making the native especially prone to infectious contamination caused by toxins. People thus influenced are easy going in use of medical drugs, hallucinatory drugs, stimulating foodstuffs, too much good living, and overindulgence of tobacco and alcoholic drinks. There is danger of insect bites, snake bites or assault from infuriated animals, or rabies smitten dogs. If conjunct with Moon, Mars or Neptune, the native is very prone to infections in general and to epidemic infections. Beside the lower emanations, there are supposedly higher influences attributed to Ras Alhague, 'though only very few people are able to attune themselves to these influences.

The fixed star was in conjunction with the MC in the birth chart of Submarine Commander GUENTER PRIEN. He had fabulous success, but later went down with his vessel.

A little girl with conjunction of this fixed star with the Ascendant was found to be suffering from sugar diabetes when she was only 9 years old (see "Direktionen", figure 79).

62. LESATH, 23°57' Sagittarius, Ypsilon Scorpii,

in close conjunction with Lambda Scorpii, both in "the Tail end of the Scorpio" (Greek for sting is lesos), has a Mars nature blended with Mercurian influence. According to tradition, Lesath in conjunction with Mars, Uranus or Saturn, Meridian or Ascendant could point to danger from wild animals, either to be ripped up by them, massacred, or torn into pieces. This medieval interpretation is no longer correct in these days. At present, a configuration involving this fixed star points to accidents, catastrophes or operations. Corresponding to the sign Scorpio, this could work out with Appendicitis, appendectomy, or removing of haemorrhoids by operation.

In the book "Direktionen" (19) page 91 an example is given, in which Sun is 22°55' in conjunction with Lesath and at the same time in opposition to Saturn, in square to MC and in the half sum Mars/Neptune. This young man, born 15.12.1943 had an appendectomy on 9.10.1959, when Uranus progressed by measure of Solar Arc, came to the opposition to Sun and Lesath.

The mother of a Thalidomide baby was born 27.6.1935 (see Pluto Book, page 111). This fixed star was in conjunction with the Meridian.

The spy HELBIG, born 28.11.1911, has in his birth chart Mercury conjunct Lesath. He gave secret information to his Eastern employer.

If associated with a benefic stellar body, and if channelled in the right direction, there is the possibility that the energy associated with the Mars nature can make for marked achievements. This is the case with the former President of the Reichsbank, SCHACHT, who had a natal conjunction of Lesath and Jupiter.

63. ETTANIN, 27°12' Sagittarius, Gamma Draconis,

in the "right Eye of the Dragon", has a Saturn-Jupiter nature with Martian influence. According to tradition, the Saturnian predominance will give a liking for solitude. This fixed star is said to tally with dishonour, downfall and loss of prestige, if in conjunction with Saturn, on the Meridian or the Horizon. If this Saturn, however, is well placed otherwise, this star is said to be helpful for mental concentration and gives a liking for philosophical and esoteric studies.

Examples of the conjunction with the Ascendant are: The English Astrologer SEPHARIAL, and the versatile Italian artist LEONARDO DA VINCI, who, in spite of his distinguished artistic flair, remained childish all his life. He had feminine features, an inclination to homosexuality and he remained "a stranger" to the people around him (20).

Ettanin was in conjunction with Mercury in the birth chart of the astronomer NEWTON.

64. BOW OF THE ARCHER, 5°41' Capricorn,

has a Jupiter-Mars nature and is specially credited with promoting mental stimuli, enterprise and a sense of justice. Natives who are influenced by this star are promoters of idealistic and humane ideas, if this fixed star is in conjunction with Ascendant, Meridian, Sun, Moon or Jupiter.

The Bow of the Archer was in conjunction with Sun in the birth charts of the actress HILDEGARD KNEF and the Chinese dictator MAO TSE TUNG, in conjunction with Venus in the cosmogram of the French painter TOULOUSE LAUTREC and of the former President of the Reichsbank H. SCHACHT, in conjunction with Jupiter in the birth configuration of the great composer BEETHOVEN and of the cosmobiologist REINHOLD EBERTIN.

The disadvantageous outcome of the Saturn conjunction can be seen in the birth chart of King BAUDOIN, who remained childless and who has a tendency towards depressive moods.

65. IN THE HEAD OF THE ARCHER, 12°46' Capricorn, Xi Sagittarii,

a nebulous star with Mars-Saturn nature and a slight Jupiter influence. Few people are supposedly ready to receive the spiritual emanations of this fixed star. A part of the Milky Way passes through the first half of the sign Capricorn, and in it are an uncountable number of star clusters and stellar nebulae, supposedly giving rise to weak eyesight in its natives.

In the examples, one can recognize the conjunction of the fixed star with the Sun in the life of poet and sculptor ERNST BARLACH, who tried to expound "the tragic situation of man between this life and the hereafter"; and in the life of the actress GERTRUD KUECKELMANN. The writer MARY LAVATER-SOLIMAN and a conjunction with Mercury. She wrote novels pointing out how people gifted with independent action make their way into a better future; aviatrix ELLY BEINHORN had a conjunction with Uranus.

66. VEGA, 14°36' Capricorn, Alpha Lyrae,

Arabic for "The Falling Eagle", principal star in constellation Lyra; is the brightest fixed star of the Northern sky and has a Venus nature with a blend of Neptune and Mercury. The Babylonians called Vega the "Star of the Queen of Life". In a good cosmic configuration, Vega is supposed to give artistic talents especially for music and acting, but also a liking for good living. With eccentric artists, this may lead to a debauched life. Tied up with Jupiter and

Venus, Vega is said to pave the way to riches and fame. However, if other influences play a part, this wealth may be lost again. High success is promised if Vega is positioned on the Horizon or Meridian. One can find Vega in corresponding aspects in birth charts of statesmen, politicians, persons of importance and influential persons with adequate aspects. In connection with Moon or Neptune, a tendency for the occult and mysticism may be given.

Vega was conjunct with the Ascendant in charts of astrologers and writers of books on this subject KARL BRANDLER-PRACHT, SINDBAD and WEISS. Also Emperor NAPOLEON III, French politician POINCARE and former King BORIS of Bulgaria had Vega conjunct with the Ascendant. Vega was conjunct with Sun in the charts of philosopher EUCKEN, and actress MARIA SCHELL, writer PAULA LUDWIG, astronomer NEWTON, and the former Bundeskanzler ADENAUER. Vega was conjunct with Mercury in the chart of the French prophet NOSTRADAMUS, and conjunct with Mars in the charts of the former General EISENHOWER and opinion pollster NOELLE-NEUMANN.

67. ALTAIR, 1°04' Aquarius, Alpha Aquilae,

in the "Eagle" also called the "Flying Eagle", has a Mars character with a blend of Mercury and Jupiter influence. Tied up with benefic stellar bodies, Altair is credited with bestowing hardiness, courage and generosity, especially if on the Ascendant. If Mercury and Moon are positioned here, this will make people as bold as brass and foolhardy, in order to assert themselves. Near the upper Meridian and in good aspect, Altair promises rise in life and high honours. The native tries with sincere conviction to reach out for his aims with utmost will power. He will avoid nothing in order to achieve them. Altair is good for the advancement of lawyers and military men.

The Moon in conjunction with Altair at the birth of Sir ARTHUR CONAN DOYLE is quite significant. Under the pseudonym SHERLOCK HOLMES, he wrote detective stories and his books reached millions of readers.

In the birth chart of Captain Graf LUCKNER, the fixed star is found on the Meridian. During the First World War, he and his ship "Sea Devil" (German: "Seeteufel") carried out wartime "piracy" with success. Later, he undertook many lecture tours taking him all over the world. It was his aim to contribute to international understanding.

68. DENEB ALGEDI, 21°06' Aquarius, Gamma Capricorni,

on the "Back of the Goat", formerly called Nashira I, has a Saturn influence made good by added Jupiter influence. Dependent on its position in the cosmogram, it will bring a life full of change. According to Arabic tradition, Deneb Algedi will make a native become a legal adviser or counsellor, and will give such a person the ability to hold a position of trust. This fixed star make for integrity and justice and gives a knowledge of man. Therefore, we see here a refining Saturn influence. This will be achieved if the radical Saturn is well placed.

We find this fixed star in conjunction with Sun in the cosmogram of one of the best known writers of this century, BERT BRECHT. His sun is also conjunct Venus. BRECHT was a radical amongst artists. He made the art of acting an almost disillusioning(Saturnian) technique contrary to the then usual "Theatre of Illusion". He made it a point to clarify how man in reality conducts himself or should react. Originally portraying life in a rough way, he turned away from bourgois society, and later became more of a critic and educator. The biggest success of his "nihilistic" (Saturnian) period of creative activity was the "Dreigroschenoper", a modification of the English "Threepenny Opera".

The Swedish seer SWEDENBORG has the same Sun conjunction with the fixed star. However, he represents quite a different type of person.

It is interesting that Deneb Algedi is to be found in three birth charts of present day Atomic physicists. It is conjunct with MC with WERNHER VON BRAUN. Rocket designer and rocket technical writer HEINZ GARTMANN has a Uranus conjunction. The physicist BURKHARD HEIM, who is known for his space-ship calculations, has a conjunction of Sun and Deneb Algedi. In GOETHE's cosmogram there is a conjunction with Uranus, as is also the case with GUENTER VON HASE, speaker of the Government of the German Federal Republic. Correspondingly, the inventor EDISON had a Mercury conjunction.

69. FOMALHAUT, 3°09' Pisces, Alpha Piscis Australis,

Alpha of the "Southern Fish", in Arabic "fom-el-hut" (= mouth of the fish), has a Mercury/Venus character with a blending of Neptune influence.

According to tradition, Fomalhaut is of quite variable effect, either very good or very bad, depending on the overall cosmic structure. It is assumed, however, that the helpful influence is the greater one, and if in conjunction with Mercury, it is said to stimulate mental capabilities and promise success as a writer or scientist. On the Ascendant and in good aspect, tradition has it that this fixed star will make for "fame" and a name "remembered for ever". In conjunction with Venus, there will be advantages in artistic pursuits. A conjunction with Jupiter or on the Meridian will bring favour from dignitaries of the church. Tied up with either Sun or Moon, the influence of Fomalhaut is said to be quite marked.

AUGUST BEBEL, co-founder and first leader of the German Social Democrats, had Fomalhaut in conjunction with Mercury, and quite nearby his Sun in half sum of (revolutionary) Uranus/Jupiter (success). He was born into a poor family, but he became the most successful agitator of his party and when only 27 years old, he already had a seat in the Reichstag.

This fixed star also was conjunct with Mercury square Mars when the Yugoslav politician ALEKSANDER RANKOWIC was born. When only 19 years old, he became secretary of the Yugoslav Communist Party, and had to spend several years in hard labor after having been caught distributing illegal handbills. He, together with TITO, directed the counter movement and he is thought to be TITO's political heir.

The painter KARL SPITZWEG had Fomalhaut in conjunction with Mars at his birth, and Mars was in the "artist axis" Venus/Mars and in the "success axis" Jupiter/Pluto.

But the poet GEORG TRAKL had a quite negative outcome of the Mars conjunction. There also was a tie up with Saturn at the same time, however, as well as Pluto and the half sum Moon/Neptune (drug craving). He became addicted to morphine and ended his life by suicide.

The physician and successful food reformer Dr. BIRCHER-BENNER had Jupiter conjunct Fomalhaut at his birth.

NOELLE-NEUMANN, world renowned opinion pollster has Fomalhaut conjunct MC. This is quite significant in her building up of a large research institution out of a very small beginning.

70. DENEB IN THE SWAN, 4°46' Pisces, Alpha Cygni,

is a large star "in the Tail of the Swan", and corresponds to a combination of Mercury and Venus influence and is therefore favourable for artistic and scientific pursuits, which are carried out with the aim of gain.

To prove this, it is interesting to find several well known artists and writers, who could make good money with their art, be it by manifold activities in the acting art or by achieving record sales of their printed works.

HEDWIG COURTS-MAHLER, the novelist, gained extraordinary success. In fact 27 million of her books were printed. However she is not considered a literary figure in the classical sense, since her writings are considered trashy "sob stuff". At her birth, Deneb was in conjunction with Mercury, and when she had her first success as a writer, Jupiter, advanced by measure of Solar Arc, passed over Mercury and the fixed star.

The popular author ERICH KAESTNER saw some of his books sell numbers of copies of several hundreds of thousands. The complete number of his sales may reach millions. In his natal chart, Sun was in conjunction with Deneb and in half sums Mercury/Uranus and Venus/Mars.

SELMA LAGERLOEF, the Swedish writer, and the Swiss HERMANN HESSE, recorded wonderful sales of their books. Both had at their birth the Lunar North Node in conjunction with Deneb. This is significant for their many contacts with millions of readers.

The fixed star is situated on the MC in the chart of popular film actor HANS SOEHNKER.

Extraordinary success, 'though in a quite different sector, was the case with HEINRICH SCHLIEMANN. In spite of many difficulties, he became, in a short time, a successful businessman. He made himself a name as one of the best known archaeologists by his excavations of Troy in Greece. In his birth chart, Venus is in conjunction with Deneb.

Further examples: Conjunct with Ascendant - the late astrological writer and editor ALEXANDER BETHOR, founder and editor of the first German astrological journal "Zodiakus". Conjunct with Sun - the American President GEORGE WASHINGTON. Conjunct with Moon - the Italian painter LEONARDO DA VINCI.

71. ACHERNAR, 14°35' Pisces, Alpha Eridani,

on "the End of the Eridanus River", possesses a Jupiter nature with a dash of Mars-Uranus character. Well placed, it promises happiness and success by giving good morals, faithful adherence to one's religious beliefs, or a philosophical inclination. According to tradition, Achernar is credited with bestowing high offices in the church, especially if conjunct with Jupiter. Cardinal FAULHABER of Munich has Sun conjunct Achernar.

When the theosophist H. P. BLAVATSKY was born, it is thought that MC was conjunct Achernar and in "success axis" Jupiter/Pluto. (Her birth minute is recorded, and we therefore assume that her birth time has been rectified.)

The King of poets J. W. GOETHE had his natal Moon conjunct with this fixed star.

A conjunction with the MC is also found with ALBERT EINSTEIN who, as atom scientist-researcher-pioneer, made an extraordinary impact both in science circles and on the public at large.

RAINIER III 'though born as a Prince, had inherited a tiny State, just the size of a middle-sized German town. He was able to make this into an attraction for visitors from all over the world and to make himself a respected man.

The Spanish President of State, General FRANCO, has in his birth chart Achernar in conjunction with Mars, being equivalent to his dictatorial regime.

Again, Mars in conjunction with the fixed star is found in the map of stigmata THERESE NEUMANN, but at the same time bound up by aspects with Saturn and Pluto, resulting in this woman, by her continued re-living of the passion of Christ, i.e. by actual suffering, becoming known world wide and exercising an influence in accordance with this.

72. MARKAB, 22°49' Pisces, Alpha Pegasi,

is "the Saddle of Pegasus" and has a Mercury-Mars nature. Tradition has it that, in conjunction with Mars, Uranus or Saturn, this will bring dangers by fire, weapon or explosion.

Tied up with propitious stellar bodies, Markab is said to influence above all the spiritual and mental nature, to give a good head for figures, intellectual alertness, mental powers in general and, last but not least, the ability to further propaganda activity, if at the same time other relevant aspects are also present.

The long lasting German Chancellor Dr. ADENAUER, had in his birth chart this fixed star in conjunction with Mars and also in square to MC and in half sums Mercury/Jupiter and Moon/Pluto. All together, a correct interpretation for his activity as a successful statesman.

The "Swedish Onassis", AXEL BROSTROEM, had unusual success with his giant shipping firm and as a ship builder. At his birth, Jupiter was united with Markab in Sun/Mars on the Ascendant.

On the other hand a conjunction with the MC can be really of grave disadvantage as the following three examples will show: Young girl, born 7.6.1932, went cycling with a basket full of flowers on 30.6. 1950. A truck travelling in the opposite direction grazed her basket. She fell down, and the rear wheel of the truck crushed her head. This fatal accident was triggered off by directed Neptune and Pluto (see "Direktionen", page 87).

Male, born 15.12.1943 - At his birth MC was conjunct with Markab, as well as in critical aspect to Sun and Saturn, in half sum Sun/ Saturn respectively. When Uranus s went over this point, he had to undergo an operation to have his appendix removed.

In the natal chart of a female, born 13.3.1903, the fixed star is in conjunction with MC and Sun, in opposition to Moon in semisquare to Saturn. When Neptune s passed over this complex, she got an infection and died at an early age from Multiple Sclerosis.

73. SCHEAT, 28°43' Pisces, Beta Pegasi,

"Shoulder of Pegasus" is a fixed star of accentuated Saturnian character. Tied up with "malefics", this could lead the native to lose his life in catastrophes, such as floods, shipwreck, mining accidents, aeroplane accidents, or maybe by suicide. In a mundane chart, with transiting MC in conjunction or transiting AS in conjunction - an accident or catastrophe may happen. The first atom explosion 18.7.1945 saw Scheat in conjunction with transiting MC and semisquare to Saturn. When the severe Alaskan earthquake occured on 27/28.3.1964, Scheat was conjunct with Mars square MC and opposition Ascendant.

On the other hand, it is possible for a positive influence to emanate from Scheat, but only for some people and having an effect on their mental creativity, if these people are ready to receive such an inflow. This is the situation with GOETHE and NIETZSCHE. Both had a Jupiter conjunction with this fixed star. The French poet VICTOR HUGO and the anthroposophist Dr. RUDOLF STEINER had their birth charts showing Mercury conjunct Scheat. MAX VALIER, inventor of the rocket-car, who also tried to further space exploration, had Scheat conjunct MC square Jupiter. The same configuration is found with the writer ERNST JUENGER, as well as with Television personality PETRA KRAUSE, who, in spite of severe handicaps, remained courageous and even married, although tied to a wheel chair.

The Duchess of WINDSOR had one of the disadvantages effects of Scheat in her life. At her birth, Sun was in conjunction with Scheat, and in an unhelpful connection with Saturn. Because of her, the then King EDWARD VII abdicated. Only recognized at a later age, the Duchess of Windsor was for decades not even allowed to enter the British Isles, and she had to live with her husband in France.

Mass murderer SEEFELD had the fixed star in conjunction with Ascendant and in square to Saturn.

As seen from these examples, the conclusion may be drawn that the fixed star Scheat has a bad effect only if tied up with "malefics".

There was a total eclipse of the Sun on 22 September 1968 and the fixed star Scheat was in conjunction with Uranus and Sun. In the diary of REINHOLD EBERTIN, the following events are recorded; events which manifest the calamitous effect very well: Very serious storms in Switzerland, many valleys were flooded for 24 hours, and a landslide near Zurich interrupted traffic on the Autobahn. In Turkey, heavy storms and hail put several villages under water. In Italy, near Cirano, a church collapsed. The Upper Rhine river showed the highest water-mark in flood disaster history. In a railway accident near Djakarta, 60 lives were lost. In Mexico as well as in Montevideo, there were serious street riots. There was a commotion and hand fighting outside the church of St. Paul in Frankfurt ("Paulskirche" is the name) on the occasion of the conferring of the peace prize. In Moscow, there was a power struggle. A dead baby was conveyed by mail to the British Queen as a protest gesture against Britain sending weapons to Biafra. In addition, there were many smaller accidents with lives lost and people injured.

RADIO WAVES AND STRUCTURAL ELEMENTS OF FIXED STARS

Dr. THEODOR LANDSCHEIDT published in 1963 an "Ephemeris of Fixed Stars connected by structural elements". The author chose fixed stars which are in aspects of either 0°, 45°, 90°, 135° and 180°. The resulting following groups are:

FS 5 A figure consisting of five fixed stars:
Sirius, Vega, Canopus, Regulus, Bungula

FS 4 A figure consisting of four fixed stars:
Antares, Aldebaran, Spica, Arcturus

FS 2 A figure consisting of two fixed stars:
Rigel and Altair.

The opposite figure has been drawn to show the structural elements. In general, only the conjunction of a fixed star and a natal stellar body (planets, Sun or Moon) is of value. Dr. LANDSCHEIDT, as well as other researchers, came to the conclusion by experience that the combination figure thus shown is of marked effectiveness. In the 90° circle, put around the 360°, there is shown how in this way there appears a kind of massing area e.g. in 14° Sirius, Canopus and Vega all meet, whilst opposite, Regulus and Bungula coincide. In about 23°-24° in FS 4 are Spica and Arcturus together and opposite these are found Antares and Aldebaran. In FS 2, Altair and Rigel are positioned opposite.

Besides this, Dr. LANDSCHEIDT introduced Apex, the target point of the Sun movement. The position for this Sun Apex for about 1970 is 2°02' Capricorn, and the position for the Galactic Centre is 26°27' Sagittarius (6).

The same author also pointed to new ways to incorporate radio waves. In addition, he has compiled an ephemeris. Please see further in the Bibliography.

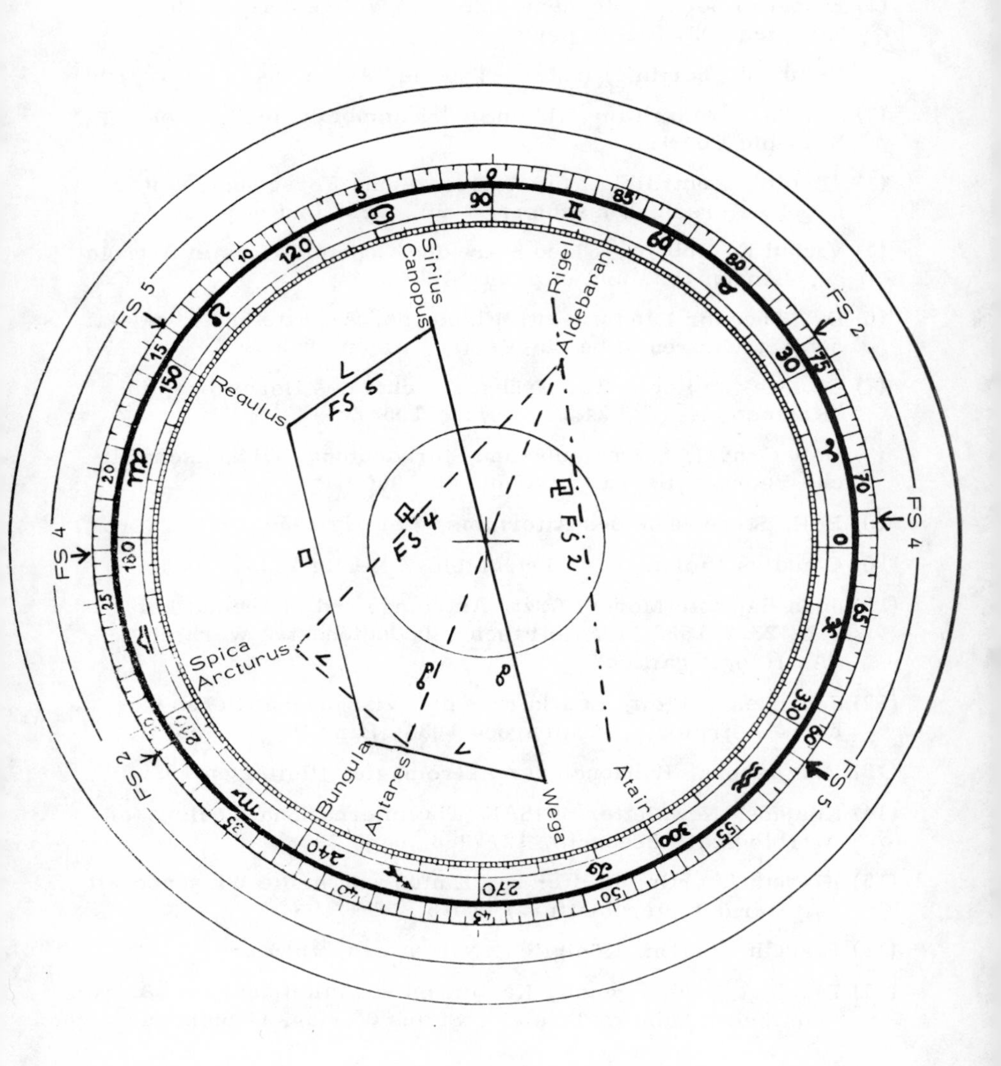

Sirius
Canopus
Rigel
Aldebaran
Regulus
FS 5
FS 4
FS 2
Spica
Arcturus
Bungula
Antares
Wega
Atair
FS 5
FS 2
FS 4
FS 4
FS 2
FS 5
0
90
120
150
180
240
270
300
330
0
30
60
5
10
15
20
25
30
35
40
43
50
55
60
65
70
75
80
85

BIBLIOGRAPHY AND OTHER REMARKS

(1) Elsbeth Ebertin, Sternenwandel und Weltgeschehen 1928, Kempten 1927 (out of print)

(2) Reinhold Ebertin, Fixstern-Tabelle, Aalen 1947 (out of print)

(3) Journal "Mensch im All", now "Kosmobiologie", editor Reinhold Ebertin

(4) "Zenit", Zentralblatt für Astrologische Forschung, editor Dr. H. Korsch, Düsseldorf, 1930 - 1938 (out of print)

(5) Vivian E. Robson, Fixed Stars and Constellations in Astrology, London 1923 (out of print)

(6) Dr. Theodor Landscheidt, Fixsterne, Aspekte und galaktische Strukturen, Ebertin-Verlag, Aalen 1965

(7) Dr. Walter Koch, Zur Frühgeschichte des Horoskops, in: "Kosmobiologie", 24. J., 10/11, 1958

(8) Boll-Gundel, Sternglaube und Sterndeutung, Wissenschaftliche Buchgesellschaft, Darmstadt 1966

(9) Boll, Sternkunde des Altertums, Leipzig 1950

(10) Claudius Ptolemaeus, Tetrabiblos, Berlin 1923

(11) Jean Baptiste Morin, Arzt, Astrologe und Mathematiker, geb. 23.2.1583 in Villefranche, bedeutendstes Werk: "Astrologia gallica"

(12) Johannes Vehlow, Lehrkursus der wissenschaftlichen Geburts-Astrologie, Zeulenroda 1934, Band II

(13) Dr. Asboga, Handbuch der Astromagie, Pfullingen 1925

(14) Kosmos, Newsletter of ISAR (The International Society for Astrological Research), 12/1968, page 29

(15) Elsbeth Ebertin, Blätter zur Einführung in die Wissenschaft der Sterne ("Sternblätter"), 1915

(16) Elsbeth Ebertin, Königliche Nativitäten, 1916

(17) Dr. Inge Koch-Egenolf, Kartomantie, Anleitung zum "Astromantischen Schicksalsspiel", Sirius-Verlag, Göppingen

(18) Reinhold Ebertin, Pluto-Entsprechungen zum Weltgeschehen und zum Menschenleben, Aalen 1965

(19) Reinhold Ebertin, Direktionen - Mitgestalter des Schicksals, Aalen 1967

(20) Lange-Eichbaum-Kurth, Genie, Irrsinn und Ruhm, Genie-Mythos und Patographie des Genies, München 1967

(21) Dr. Theodor Landscheidt, Die astronomischen Radio-Quellen und ihre Auswirkung auf das solare und irdische Geschehen, Hamburg 1964

(22) Dr. Theodor Landscheidt, Ephemeride der durch Strukturen verbundenen Fixsterne nebst einer Einführung in den Gebrauch, Aalen 1963 (out of print).

EXPLANATION OF THE FIXED STAR TABLE

The table "Latitudinal Positions of the Fixed Stars" shows the position of each fixed star in its Zodiacal sign, and then its brightness and magnitude. Fixed stars of first magnitude are considered as the important ones. The latitude of the fixed stars is measured by its position on the ecliptic. In general, it is the custom that only longitudinal positions are taken into consideration, when fixed stars are mentioned. These positions are given for 1900, 1920, 1937 and 1950. As seen from the resulting difference, the movement of the fixed stars is a minute one. The yearly motion is shown on the end of each line. Besides this movement of the star itself, there is also a so-called apparent movement on account of the equinoxes, which results in about 72 years = 1°, or in other words, 0°0'50" in one year. This movement is in the opposite direction to the Sun movement. It therefore appears as though the fixed stars are moving, but this is not so. The fixed stars are stationary, the Spring and Autumn points are the ones which are the originators of this apparent movement. There is a separate table to calculate this motion, e.g. to calculate a fixed star for 1890, one has to make a deduction of 8'23" from the position of 1900. To calculate a position for 1970, i.e. 20 years later than 1950, one has to add 16'45" to the given positions in the table.

With the different precision, there is also a different latitude and declination. But these changes are so minute that they really can be overlooked. It is far more important to note whether the fixed star is near the ecliptic, near which all other stellar bodies make their track (as seen from earth) and how this will therefore make a "genuine" conjunction, or whether the fixed star is quite distant from the ecliptic, as e.g. the Pole Star, thus making a large difference in latitude.

Latitude, Rectascension and declination of the fixed stars and their yearly motion are shown in another table. The latitude of a fixed star signifies the distance from the ecliptic, which is the declination of the distance from the celestial equator. The fixed star Praesepe has a latitude of only 1°35' N., i.e. it is only 1°33' away from the ecliptic in a Northerly direction. The distance from the celestial equator however, (i.e. the declination) is 20°15' North. It is possible that S (South) and N (North) are differing in latitude and declination. Sertan has a latitude of 4°54' South, i.e. it is a distance of 4°54' from the equator and South of the path of the Sun. But Sertan has a Northerly declination of 12°19' as it is such a distance away from the equator in a Northerly direction. In order to spell out these various differences, a special Fixed Star Table has been published besides this book. In this, one can see positions of the fixed stars in a figure, arranged according to longitude, latitude and declination. As well as this, the "nature of the fixed star", or the corresponding character as compared with a planet, has been given. One can therefore use this table as a practical help in order to avoid some of the mistakes that are possible.

Note:

Dr. W. KOCH had looked through the manuscript briefly. On the last day before the printing plates were finished, he wrote to us: "Alpha Centauri" is the bright star on the front part of the foot which is put forward, Rigil Kent Beta Centauri is somewhat less bright than the former and is on the second foot on the front part. Algena Rigil Kent means 'The Foot of the Centaur'. The name Bungula is supposed not to be used any more in future. Toliman is the alternative name for Alpha Centauri, but can be misunderstood and therefore Alpha Centauri should not be called Toliman any more.

According to KLEPESTA, Toliman means 'Shoot of the Vine'. This is an indication of the Rod of Tyrsus mentioned by PTOLEMY, containing 4 stars and mainly Psi. In recent times, this Rod of Tyrsus has been changed to a spear pointing towards the wolf. Therefore, properly interpretated, Toliman means 'main star of the Rod of Tyrsus'. Its latitude is approx. - 22°, the latitude of Alpha Centauri approx. - 41°. These values are noted down by PTOLEMY. The difference of latitude is therefore about 19°." - It is the sincere wish of the authors to establish a correct basis for research into fixed stars. Every hint or correction will be appreciated at any time.

POSITION IN LONGITUDE OF FIXED STARS

No.	Name of fixed star		1900	1920	1937	1950	Motion	Character
1	β Ceti	Deneb Kaitos (2)	1° 09' ♈	1° 28'	1° 41' 46"	1° 51' 00"	+ 45,1	♄
2	γ Pegasi	Algenib (2)	7 48	8 02	8 17 01	8 31	46,3	♂☿
3	α Andromedae	Sirrah (2)	12 57	13 11	13 25 55	13 40	46,5	♃♀
4	ζ Ceti	Batan Kaitos (3)	20 30	20 50	21 03	21 17	44,4	♄
5	β Andromedae	Mirach (2)	29 01	29 17	29 31 36	29 46	50,2	♀
6	ο Ceti	Mira (2–3)	0 07 ♉	0 25	0 39	0 50	51,6	♄♃
7	β Arietis	El-Scheratain (3)	2 35	2 51	3 05 24	3 17	49,6	♂♄
8	α Arietis	El - Nath (2)	6 17	6 35	6 46 53	6 59	48,0	♂♄
9	α Cassiopeiae	Schedir (1)	6 25	6 42	6 54 36	7 07	50,7	♄♀
10	γ Andromedae	Alamak (2)	12 52	13 07	13 20 53	13 34	55,0	♀♂
11	α Ceti	Menkar (2)	12 56	13 12	13 26 23	13 38	46,9	♄
12	γ Eridani	Zanrak (3)	22 08	22 25	22 36	22 50	50,4	♄
13	β Persei	Algol (2–4)	24 47	25 03	25 17 22	25 28	58,6	♄♂♅⊖
14	η Tauri	Alcyone-Plejades (3)	28 36	28 52	29 06 46	26 19	53,4	☽♂
15	γ Tauri	Hyades (4)	4 26 ♊	4 41	4 55 26	4–6°	51,1	♂♆♅
16	ε Tauri	Northern Bull Eye (4)	7 05	7 22	7 36	7° 47'		♀☽
17	α Tauri	Aldebaran (1)	8 24 L	8 40	8 54 31	9 05 25 L	50,0	♂
18	β Orionis	Rigel (1)	15 28	15 43	15 56 56	16 07 50 L	43,2	♂♃
19	γ Orionis	Bellatrix (2)	19 34	19 50	20 03 20	20 16	48,3	♂☿
20	Aurigae	Capella (1)	20 26	20 44	20 56 39	21 10	66,4	☿♂
21	α Columbae	Phact (3)	20 47	21 03	21 17 20	21 29	32,5	☿♀
22	α Ursae min.	Pole Star (2)	27 12	27 27	27 38 57	27 54	50,6	♄☉♀
23	α Orionis	Beteigeuze (1)	27 22	27 38	27 52 28	28 04	48,7	♂☿
24	Aurigae	Menkalinam (2)	28 30	28 48	29 01 53	29 13	66,0	♃♂♀

No.	Name of fixed star		1900	1920	1937	1950	Motion	Character
25 γ	Geminorum	Alhena (2)	7°42' ♋	7°59'	8°13'28"	8°24'	52,0"	♀♃
26 α	Canis maior	Sirius (1)	12 45 L	12 59	13 12 49	13 23 33 L	39,6	♂♃
27 α	Carinae	Canopus (1)	13 35	13 51	14 05 22	14 16 12 L	18,3	♃♄
28 α	Geminorum	Castor (2)	18 51	19 08	19 21 46	19 33	18,3	☿♃
29 β	Geminorum	Pollux (2)	21 53	22 07	22 20 50	22 35	55,2	♂
30 α	Canis min.	Procyon(1)	24 28	24 41	24 54 59	25 10	47,1	♂☿
31 ε	Cancri	Praesepe (6)	5 52 ♌	6 07	6 31 12	6 34	51,6	☽♂
32 γ	Cancri	North Aselli(5)	6 08	6 25	6 39	6 50	52,3	☉♂
33 δ	Cancri	South Aselli(4)	7 18	7 36	7 50	8 01	51,1	☉♂
34 β	Ursae min.	Kochab (2)	11 42	11 49	12 03	12 24		♂
35 α	Cancri	Sertan (4)	12 14	12 31	12 45	12 56	49,2	♂♄
36 α	Ursae maior	Dubhe (2)	13 45	14 02	14 18 13	14 27	45.0	♂
37 β	Ursae maior	Merak (2)	17 59	18 16	18 32 48	18 41	46.2	♂
38 ε	Leonis	Ras Elased Austr. (3)	19 18	19 35	19 49 30	20 00	51,1	♂♄
39 α	Hydrae	Alphard (1)	25 54	26 10	26 24 11	26 36	44,2	♄♀♆
40 α	Leonis	Regulus (1)	28 27 L	28 43	28 57 13	29 08 L	48,0	♃♂
41 γ	Ursae maior	Phacht (2)	28 59	29 16	29 29	29 41		♂♅♆
42 ε	Ursae maior	Alioth (2)	7 27 ♍	7 44	8 02 19	8 09	51,4	♂
43 δ	Leonis	Zosma(3)	9 54	10 12	10 25 53	10 35	47,8	♄♀
44 ζ	Ursae maior	Mizar(2)	14 10	14 27	14 48 05	14 52	52,4	♂
45 β	Leonis	Denebola (2)	20 15	20 30	20 44 36	20 57	45,9	♅(♀♄)
46 η	Ursae maior	Benetnash(2)	25 26	25 43	26 02 42	26 08	50,6	♂♅♄
47								
47 ε	Virginis	Vindemiatrix (3)	8 34 ♎	8 50	9 03 46	9 16	44,8	♄☿
48 δ	Corvi	Algorab)(2)	12 03	12 20	12 34	12 45	46,5	♂♄
49 α	Virginis	Spica (1)	22 26 L	22 43	22 57 44	23 08 36 L	47,4	♀♂
50 α	Bootis	Arcturus (1)	22 50	23 07	23 21 15	23 32	41,1	♃♂

POSITION IN LONGITUDE OF FIXED STARS III

No.	Name of fixed star		1900	1920	1937	1950	Motion	Character
51 β	Crucis	Southern Cross (1)	10° 29' ♏	10° 46'	10° 59' 56"	11° 11'	49,8"	♃
52 α	Coronae	Gemma (2)	10 50	11 10	11 24 26	11 32	38,1	♀☿
53 α	Librae	Southern Scales (2)	13 42	13 58	14 12 15	14 23	49,6	♂♄
54 β	Librae	Northern Scales (2)	17 58	18 15	18 29 33	18 40	48,4	♃☿
55 α	Serpentis	Unuk (2)	20 41	20 56	21 11 27	21 23	44,2	♂♄
56 β	Centauri	Beta Centauri (1)	22 24	22 43	22 55 01	23 06	63,1	♀♃
57 α	Centauri	Bungula (1)	28 13 L	28 28	28 41 15	28 51 04 L	45,0 L	♀♃
58 δ	Scorpii	Akrab (3)	1 09 ♐	1 27	1 41 30	1 51	53,1	♂♄
59 α	Scorpii	Antares (1)	8 22 L	8 39	8 52 57	9 03 51 L	50,o L	♂☿♃♄
60 α	Herculis	Ras Algethi (3-4)	14 45	15 02	15 16 16	15 27	50,3	♂♀
61 α	Ophiuchi	Ras Alhague (2)	21 00	21 20	21 33 57	21 42	41,7	♄♀
62 γ	Scorpii	Lesath (2)	23 15	23 32	23 07 56 ?	23 57	48,6	♂☿
63 γ	Draconis	Ettanin (2)	26 30	26 47		27 12		♄♃♂
64 λ	Sagittarii	Bow of the Archer (3)	4 59 ♑	5 16		5 41		♃♂
65 ξ	Sagittarii (4)	Head of the Archer (4)	12 04	12 21	12 34 14	12 46	49,4	♂☉♃
66 α	Lyrae	Vega (1)	13 55 L	14 12	14 25 59	14 36 42 L	43,9	♀☿♆
67 α	Aquilae	Altair (1)	0 20 ♒	0 39	0 53 08	1 04 13 L	43,9	☿♃
68 γ	Capricorni	Deneb Algedi (4)	20 23	20 40	20 54 27	21 06	28,0	♄♃
69 α	Piscis austr.	Fomalhaut (1)	2 27 ♓	2 44	2 58 20	3 09	49,8	☿♀♆
70 α	Cygni	Deneb in the Swan (2)	4 04	4 14	4 27 46	4 46	30,6	☿♀
71 α	Eridani	Achernar (1)	13 53	14 10	14 24 58	14 35	33,6	♃♂♅
72 α	Pegasi	Markab (2)	22 08	22 22	22 36 29	22 49	44,8	☿♂
73 β	Pegasi	Scheat (2)	28 01	28 15	28 29 41	28 43	43,6	♄

TABLE OF PRECISION OF FIXED STARS

Year	Movement	Year	Movement	Year	Movement
1	0° 0' 50"	34	0° 28' 29"	68	0° 56' 57"
2	0 1 42	35	0 29 19	69	0 57 47
3	0 2 31	36	0 30 09	70	0 58 38
4	0 3 21	37	0 30 59	71	0 59 28
5	0 4 11	38	0 31 50	72	1 00 18
6	0 5 02	39	0 32 40	73	1 01 08
7	0 5 52	40	0 33 30	74	1 01 59
8	0 6 42	41	0 34 20	75	1 02 49
9	0 7 32	42	0 35 11	76	1 03 39
10	0 8 23	43	0 36 01	77	1 04 29
11	0 9 13	44	0 36 51	78	1 05 20
12	0 10 05	45	0 37 41	79	1 06 10
13	0 10 53	46	0 38 32	80	1 07 00
14	0 11 44	47	0 39 22	81	1 07 50
15	0 12 34	48	0 40 12	82	1 08 41
16	0 13 24	49	0 41 02	83	1 09 31
17	0 14 14	50	0 41 53	84	1 10 21
18	0 15 05	51	0 42 43	85	1 11 11
19	0 15 55	52	0 43 33	86	1 12 02
20	0 16 45	53	0 44 23	87	1 12 52
21	0 17 35	54	0 45 14	88	1 13 42
22	0 18 26	55	0 46 04	89	1 14 32
23	0 19 16	56	0 46 54	90	1 15 23
24	0 20 06	57	0 47 44	91	1 16 13
25	0 20 56	58	0 48 35	92	1 17 03
26	0 21 47	59	0 49 25	93	1 17 53
27	0 22 37	60	0 50 15	94	1 18 44
28	0 23 27	61	0 51 05	95	1 19 34
29	0 24 17	62	0 51 56	96	1 20 24
30	0 25 08	63	0 52 46	97	1 21 14
31	0 25 58	64	0 53 36	98	1 22 05
32	0 26 48	65	0 54 26	99	1 22 55
33	0 27 38	66	0 55 17	100	1 23 45
34	0 28 29	67	0 56 07	101	1 24 35

LATITUDE, REACTASCENSION, DECLINATION OF FIXED STARS (1920)

No.	Name	Latitude	Rectascension	Declination	Yearly motion
1	Deneb Kaitos	S 20° 46'	9° 53' 37,0"	S 18 24 52,4"	° + 19,8"
2	Algenib	N 12 36	2 16 42,6	N 14 44 19,6	20,0
3	Sirrah	N 25 41	1 03 43,8	N 28 38 35,6	19,9
4	Batan Kaitos	S 20 20	26 52 39,7	S 10 43 47,4	17,8
5	Mirach	N 25 56	16 18 42,4	N 35 11 48,4	19,1
6	Mira	S 16 07	33 34 10.	S 3 31 29.0	16,5
7	El-Scheratain	N 08 29	27 33 14,8	N 20 25 3,1	17,7
8	El-Nath	N 10 03	30 25 00	N 23 05	17,2
9	Schedir	N 9 58	30 39 53,1	N 56 07	17,1
10	Alamak	N 27 48	29 44 42,9	N 41 56 47,2	17,3
11	Menkar	S 12 35	44 31 25,6	N 3 46 35,9	14,2
12	Zanrak	S 31 53	58 20 56,0	S 13 50 05,0	10,4
13	Algol	N 22 25	45 44 21,1	N 40 38 54,4	14,0
14	Alcyone-Plejades	N 4 2	55 40 52,9	N 23 51 31,5	11,3
15	Hyades	S 5 44	63 48 34,5	N 15 26 7,6	8,8
16	Northern Bull Eye	S 2 43	65 43 10,0	N 19 00 44,0	8,2
17	Aldebaran	S 5 28	67 49 55,2	N 16 20 58,3	7,4
18	Rigel	S 31 08	77 40 23,1	S 8 17 35,2	4,3
19	Bellatrix	S 16 50	80 12 35,5	N 6 16 41,5	3,4
20	Capella	N 22 55	77 41 38,7	N 45 55 5,1	3,8
21	Phact	S 57 23	84 11 16,2	S 34 6 58,1	2,0
22	Pole Star	N 66 5	22 55 18,1	N 88 52 38,9	18,4
23	Beteigeuze	S 16 2	87 42 36,3	N 7 23 35,9	0,8
24	Menkalinam	N 21 30	88 24 54,6	N 44 56 26,7	0,5
25	Alhena	S 6 45	98 16 21,9	N 16 28 7,3	2,9
26	Sirius	S 39 35	100 24 22,0	S 16 36 19,9	4,8
27	Canopus	S 75 50	95 32 37,5	S 52 39 5,5	1,9
28	Castor	N 10 5	112 22 26,8	N 32 3 55,9	7,7
29	Pollux	N 6 40	115 6 21,0	N 28 13 14,9	8,6
30	Procyon	S 16 00	113 46 43,6	N 5 25 51,8	9,1
31	Praesepe	N 1 33	128 54 15,0	N 20 15 8,0	12,2
32	North Aselli	N 3 11	129 40 1,0	N 21 45 43,3	12,5
33	South Aselli	N 0 4	130 2 7,3	N 18 26 57,1	13,1
34	Kochab	N 56 26	222 45 5,0	N 74 39 00,0	14,7
35	Sertan	S 4 54	133 16 5,8	N 12 19 35,0	13,8
36	Dubhe	N 49 56	164 24 15,0	N 62 23 28,0	19,4
37	Merak	N 45 08	163 58 13,0	N 57 01 24,4	19,22
38	Ras Elased Austr.	N 9 50	145 04 08,2	N 24 19 28,8	16,4
39	Aphard	S 22 23	140 54 51,1	S 8 18 40,2	15,5
40	Regulus	N 0 28	151 1 42,3	N 12 21 31,3	17,5

LATITUDE, RECTASCENSION, DECLINATION OF FIXED STARS (1920)

No.	Name	Latitude	Rectascension	Declination	Yearly motion
41	Phacht	N 46° 51'	177° 09' 3,6	N 54 21 40	-20,0"
42	Alioth	N 54 11	192 24 53,0	N 56 37 32	19,6
43	Zosma	N 14 20	167 27 50,8	N 20 57 44,0	19,7
44	Mizar	N 56 32	199 59 04,0	N 55 32 40,0	19.7
45	Denebola	N 12 16	176 14 42,4	N 15 1 9,6	20,1
46	Benetnasch	N 53 56	206 01 54,0	N 49 55 02,0	18,1
47	Vindemiatrix	N 16 13	194 32 55,2	N 11 23 19,8	19,4
48	Algorab	S 12 11	186 25 50,2	S 16 14 41,0	20,1
49	Spica	S 2 3	200 14 38,4	S 10 44 39,0	18,8
50	Arcturus	N 30 47	213 0 10,6	N 19 35 54,0	18,8
51	Southern Cross	S 43 17	190 29 38,0	S 59 15 34,4	19.7
52	Gemma	N 44 20	232 49 30,3	N 26 58 59,2	12,2
53	Southern Scales	N 0 20	221 36 44,2	S 15 42 36,6	15,1
54	Northern Scales	N 8 30	228 10 29,5	S 9 5 19,2	13,4
55	Unuk	N 25 25	235 4 53,5	N 6 40 35,1	11,4
56	Beta Centauri	S 44 19	209 32 27,7	S 59 57 49,0	17,5
57	Bungula	S 42 34	218 32 18,1	S 60 30 21,8	15,0
58	Akrab	S 1 58	238 53 59,7	S 22 23 42,7	10,4
59	Antares	S 4 34	246 7 29,1	S 26 15 20,4	8,1
60	Ras Algethi	N 37 17	257 44 59,6	N 14 27 46,0	4,3
61	Ras Alhague	N 35 51	262 48 18,0	N 12 37 1,9	2,7
62	Lesath	S 14 00	261 19 48,6	S 37 14 00,0	3,0
63	Ettanin	N 74 21	268 34 27,8	N 51 30 10,6	0,5
64	Bow of the Archer	S 1 59	275 28 02,0	S 25 34 42,0	1,8
65	Head of the Archer	N 1 47	282 57 11,6	S 21 15 29,4	4,5
66	Vega	N 61 44	276 33 26,7	N 38 42 30,3	3,3
67	Altair	N 29 18	296 43 12,0	N 8 39 21,9	9,4
68	Deneb Algedi	S 2 35	325 39 24,6	S 16 29 27,6	16,3
69	Fomalhaut	S 21 08	343 18 29,5	S 30 2 47,5	19,0
70	Deneb in the Swan	N 59 55	309 40 33,7	N 44 59 37,7	12,8
71	Achernar	S 59 22	23 41 3,6	S 57 38 34,5	18,3
72	Markab	N 19 24	345 11 36,9	N 14 46 28,2	19,3
73	Scheat	N 31 8	344 58 24,3	N 27 38 54,7	19,5

REINHOLD EBERTIN

FIXED STAR TABLE

with position for 1950

EXPLANATION:

This Fixed Star Table is an appendix to the book written by EBERTIN - HOFFMANN, "Fixed Stars and their Interpretation" and is meant to facilitate quick locating of fixed star positions.

Right through the middle is the path of the heavenly equator with each sign of the Zodiac divided into degrees. The ecliptic or the Solar path moves first from the spring point (first point of Aries) towards North, is then culminating in the sign Cancer, and then moves again towards the equator, cuts through this in the Autumn point and reaches, in the sign Capricorn, the lowest position South of the equator. One can see quite easily whether a fixed star is positioned near the path of the Sun and therefore whether it is also near other stellar bodies (as seen from Earth). At the same time, one easily recognizes their distance away from the equator. It is accepted that fixed stars nearer the ecliptic have greater intensity of effect than those further away.

All positions are given for the year 1950. Precise positions for other years can be interpolated according to instructions given with the Fixed Star Table in the book, "Fixed Stars and their Interpretation". The structures of fixed stars according to Dr. LANDSCHEIDT are shown separately and distinctively. Immediately one can see by this means when a fixed star is in aspect with other fixed stars. At the same time the symbols will convey the character of the fixed stars. This table should become a valuable aid to all friends of Cosmobiology.

Reinhold Ebertin

This table was wholly produced by Ebertin-Verlag, D 7080 Aalen/ West Germany, Tulpenweg 15, Postfach 1223.

REINHOLD EBERTIN : FIXE

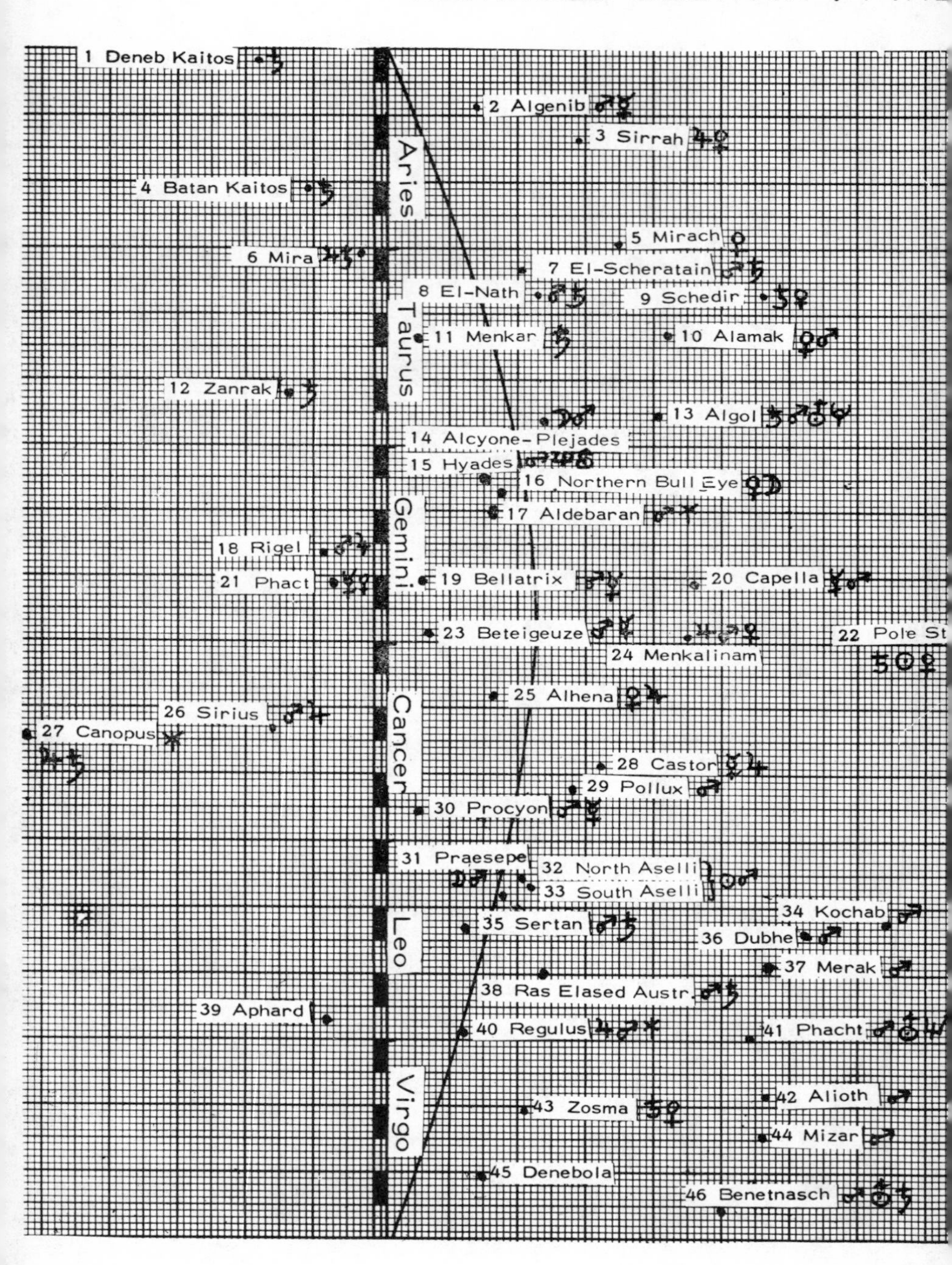

AR TABLE for 1950

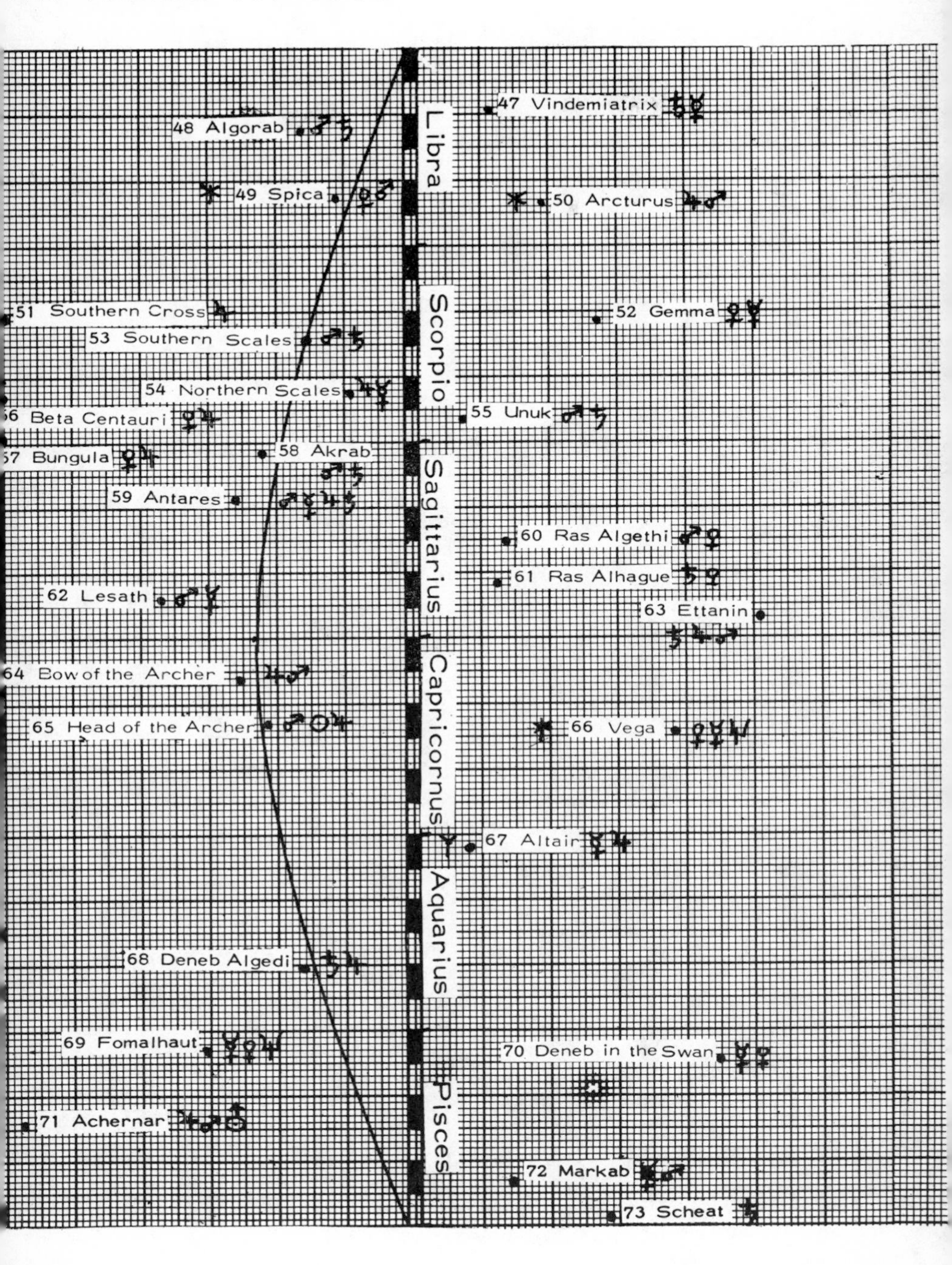